T0240525

Lecture Notes in Mathematics 2225

More information about this series at http://www.springer.com/series/304

Liguang Liu • Jie Xiao • Dachun Yang • Wen Yuan

Gaussian Capacity Analysis

Springer

Liguang Liu
School of Mathematics
Renmin University of China
Beijing, China

Jie Xiao
Department Mathematics & Statistics
Memorial University
St. John's, Newfoundland and Labrador
Canada

Dachun Yang
Laboratory of Mathematics and Complex
Systems (Ministry of Education of China)
School of Mathematical Sciences
Beijing Normal University
Beijing, China

Wen Yuan
Laboratory of Mathematics and Complex
Systems (Ministry of Education of China)
School of Mathematical Sciences
Beijing Normal University
Beijing, China

ISSN 0075-8434   ISSN 1617-9692   (electronic)
Lecture Notes in Mathematics
ISBN 978-3-319-95039-6   ISBN 978-3-319-95040-2   (eBook)
https://doi.org/10.1007/978-3-319-95040-2

Library of Congress Control Number: 2018951056

Mathematics Subject Classification: 31B15, 31B35, 42B35, 52A38, 53A07, 53C65, 60D05

This Springer imprint is published by the registered company Springer Nature Switzerland AG.
The registered company address is: Gewerbestrasse 11, 6330 Cham, Switzerland

Preface

This monograph *Gaussian Capacity Analysis* documents the functional p-capacity and its extreme BV-capacity in Gauss space with applications to tracing the Gaussian Sobolev p-space and its extreme BV-space as well as the induced geometric structure.

Recall that the Gauss space $\mathbb{G}^n$ is the $1 \le n$-dimensional Euclidean space $\mathbb{R}^n$ equipped with the standard Euclidean metric

$$|x - y| = \left(\sum_{j=1}^n (x_j - y_j)^2 \right)^{\frac{1}{2}} \quad \forall \quad (x = (x_1, \ldots, x_n), y = (y_1, \ldots, y_n)) \in \mathbb{R}^n \times \mathbb{R}^n$$

and the canonical Gaussian measure density

$$\gamma(x) := (2\pi)^{-\frac{n}{2}} e^{-\frac{|x|^2}{2}} \quad \forall \quad x \in \mathbb{R}^n.$$

Denote by dV and $dV_\gamma = \gamma dV$ the Lebesgue volume element and the Gaussian volume element respectively. Note that V_γ is a canonical probability measure, i.e.,

$$V_\gamma(\mathbb{R}^n) = \int_{\mathbb{R}^n} \gamma(x) \, dV(x) = 1.$$

The Gauss space arises from probability theory, quantum mechanics, and differential geometry; see [8, 43, 44, 45, 26, 42], and the references therein.

If $C_c^0(\mathbb{R}^n)$ represents the class of continuous functions with compact support in $\mathbb{R}^n$, then for any integer $k \ge 1$ the class $C_c^k(\mathbb{R}^n)$ consists of all k-order continuously differentiable functions with compact support in $\mathbb{R}^n$. Given $p \in [1, \infty)$. It was proved by Pisier [47, Corollary 2.4] that

$$(\star) \quad \left(\int_{\mathbb{R}^n} \left| f - \int_{\mathbb{R}^n} f \, dV_\gamma \right|^p dV_\gamma \right)^{\frac{1}{p}} \le C_{p,n} \left(\int_{\mathbb{R}^n} |\nabla f|^p \, dV_\gamma \right)^{\frac{1}{p}} \quad \forall f \in C_c^1(\mathbb{R}^n),$$

where $C_{p,n}$ is a positive constant depending only on n and p; see also Ledoux [34, pp. 212–213]. As is well known, this last inequality can be regarded as the Gaussian Poincaré p-inequality .

Here, it is worth mentioning that $(\star)$-like inequalities over $\mathbb{G}^n$ were investigated by many authors from different perspectives such as probability, partial differential equation, complex analysis, and convex geometry. In particular, when $p = 1$, via a probability method Ledoux [34, p. 279] obtained $(\star)$ with a sharp constant $C_{p,n} = \sqrt{\frac{\pi}{2}}$. When $p = 2$, the Gaussian Poincaré 2-inequality is closely related to the first nontrivial eigenvalue of the following problem (see [12]):

$$\begin{cases} -\mathrm{div}\left(\left(\exp(-\frac{|x|^2}{2}) \right) \nabla u(x) \right) = \lambda \left(\exp(-\frac{|x|^2}{2}) \right) u(x) & \text{in } \Omega; \\ \frac{\partial u}{\partial \nu} = 0 & \text{on } \partial\Omega, \end{cases}$$

where Ω is a bounded convex domain of $\mathbb{R}^n$ and ν stands for the outward normal to $\partial\Omega$. When $p \in [2, \infty)$, $(\star)$ was also derived by Zeng [58, Proposition 2.1] with $C_{p,n} = C\sqrt{p}$.

Observe that an equivalent statement of $(\star)$ is as follows: there exists a positive constant $C_{p,n}$ such that

$$(\star\star) \quad \left(\int_{\mathbb{R}^n} |f|^p \, dV_\gamma\right)^{\frac{1}{p}} \le C_{p,n} \left(\left(\int_{\mathbb{R}^n} |\nabla f|^p \, dV_\gamma\right)^{\frac{1}{p}} + \left|\int_{\mathbb{R}^n} f \, dV_\gamma\right|\right) \ \forall f \in C_c^1(\mathbb{R}^n).$$

So, it is natural to ask the following restriction/trace question:

Given $q \in (0, \infty)$, *for which nonnegative Borel measure* μ *on* $\mathbb{R}^n$ *does the inequality*

$$(\star\star\star) \quad \left(\int_{\mathbb{R}^n} |f|^q \, d\mu\right)^{\frac{1}{q}} \le C \left(\left(\int_{\mathbb{R}^n} |\nabla f|^p \, dV_\gamma\right)^{\frac{1}{p}} + \left|\int_{\mathbb{R}^n} f \, dV_\gamma\right|\right)$$

hold uniformly over suitable functions f *with a positive constant* C *independent of* f?

In order to answer this question, we introduce and study the so-called Gaussian p-capacity, thereby completing this monograph of five chapters.

▶ After exploring the basic structure of a Gaussian Sobolev p-space in Chapter 1, it is easy to observe that the right sides of $(\star\star)$ and $(\star\star\star)$ are equivalent to the Gaussian Sobolev norm $\| \cdot \|_{W^{1,p}(\mathbb{G}^n)}$ induced by the right side of $(\star)$.

▶ Meanwhile, the left side of $(\star)$ (coupled with [36, 37]) leads to Chapter 2 - an investigation of the Gaussian Campanato class and its relationship to the Morrey or John-Nirenberg or Lipschitz class on $\mathbb{G}^n$.

▶ By the phenomena appeared in Chapters 1–2, it comes quite natural in Chapter 3 that the Gaussian p-capacity of a compact set K can be defined in the following way:

$$\inf \left\{ \|f\|_{W^{1,p}(\mathbb{G}^n)}^p : f \in C_c^1(\mathbb{R}^n) \text{ with } f \ge 1 \text{ on } K \right\}.$$

Accordingly, the fundamental properties of this new capacity will be presented over there.

▶ Via this new capacity, we shall later in Chapter 4 characterize such a nonnegative Radon measure μ on $\mathbb{R}^n$ obeying $(\star\star\star)$.

▶ As in [12, 27, 53], the Gaussian 1-capacity of fundamental importance can be characterized by the Gaussian Minkowski content of the boundary of a set - consequently - not only an equivalence between the Cheeger's isoperimetric inequality on $\mathbb{G}^n$ and the Gaussian 1-Poincaré inequality can be discovered to reprove ($\star$) under $p = 1$, but also Ehrhard's inequality and its immediate product - the Gaussian isoperimetry can be induced; and moreover the Gaussian ∞-capacity will be discussed as a dual form of the Gaussian 1-capacity – see Chapter 5.

▶ As the weakest formulation of the Gaussian 1-capacity, the Gaussian BV-capacity is addressed in Chapter 6 which is from a suitable modification of [53] and its main reference [27].

Briefly and historically speaking, the concept of a functional capacity is originated from electrostatics. The study of an electrical capacitance was started by Wiener in 1924 and has achieved a great development from then on; see, e.g., Pólya [48], Choquet [14], and Shubin [50]. The functional capacities are of fundamental importance in various branches of mathematics such as analysis, geometry, mathematical physics, partial differential equations, and probability theory; see, e.g., [2, 5, 10, 20, 41, 18, 59, 15, 16, 27, 30, 31, 54, 55, 21, 22, 38, 7, 29, 19]. Needless to say, this monograph is a new development of the geometric potential analysis based on the Gauss space and will be definitely useful for researchers working in the above-mentioned fields and their relatives.

Liguang Liu was partially supported by NNSF of China Grant # 11771446 & #11761131002; Jie Xiao was partially supported by NSERC of Canada Grant # 20171864; Dachun Yang was partially supported by NNSF of China Grant # 11571039 & #11671185 & #11761131002 & #11726621; Wen Yuan was partially supported by NNSF of China Grant # 11471042 & #11871100 & #11761131002 & #11726621.

Last but not least, we would like to point out that the following conventions will be used throughout this monograph.

▶ $\mathbb{N} := \{1, 2, 3, \ldots\}$ & $\mathbb{Z} = \{0, \pm 1, \pm 2, \ldots\}$;

▶ For any set $E \subset \mathbb{R}^n$ denote by E°, $E^c = \mathbb{R}^n \setminus E$, and 1_E the interior, the complement, and indicator of E respectively;

▶ For $k + 1 \in \mathbb{N} \cup \{\infty\}$ the symbol $C_c^k(\mathbb{R}^n)$ is the class of all functions $f : \mathbb{R}^n \to \mathbb{R}$ with k continuous partial derivatives (written as $f \in C^k(\mathbb{R}^n)$) and compact support, and

$$\begin{cases} C_c^k(\mathbb{R}^n; \mathbb{R}^m) := \underbrace{C_c^k(\mathbb{R}^n) \times \cdots \times C_c^k(\mathbb{R}^n)}_{m \text{ copies}}; \\[2mm] C^k(\mathbb{R}^n; \mathbb{R}^m) := \underbrace{C^k(\mathbb{R}^n) \times \cdots \times C^k(\mathbb{R}^n)}_{m \text{ copies}}; \end{cases}$$

▶ C represents a positive constant depending only on the main parameters involved. Occasionally, $C_{\alpha, \beta, \ldots}$ indicates that C depends on the parameters $\alpha, \beta, \ldots$;

► Given any two nonnegative quantities X and Y, the symbol $X \simeq Y$ means $X \leq CY$ (denoted by $X \lesssim Y$) and $C^{-1}X \geq Y$ (denoted by $X \gtrsim Y$) for a positive constant C.

Beijing, China   Liguang Liu
St. John's, Canada   Jie Xiao
Beijing, China   Dachun Yang
Beijing, China   Wen Yuan
December 2017–March 2018

Contents

Chapter 1
Gaussian Sobolev p-Space

In this chapter we are motivated by the right side of $(\star)$ to give an exploration of the fundamental characteristics of the Gauss-Sobolev spaces which will be used to introduce the Gaussian Sobolev p-capacity later in Chapter 3.

1.1 Definition and Approximation of $W^{1,p}(\mathbb{G}^n)$

For $1 \le p < \infty$ and all measurable functions f on $\mathbb{R}^n$, we define the norm

$$\|f\|_{L^p(\mathbb{G}^n)} := \left(\int_{\mathbb{R}^n} |f|^p \, dV_\gamma \right)^{\frac{1}{p}}.$$

If $p = \infty$, then we define

$$\|f\|_{L^\infty(\mathbb{G}^n)} := \inf \left\{ t : |f(x)| \le t \ \ a.e. \, x \in \mathbb{R}^n \right\}.$$

For $1 \le p \le \infty$ define $L^p(\mathbb{G}^n)$ to be the space of all measurable functions f on $\mathbb{R}^n$ such that $\|f\|_{L^p(\mathbb{G}^n)} < \infty$. Observe that $L^\infty(\mathbb{G}^n)$ coincides with the Lebesgue space $L^\infty(\mathbb{R}^n)$.

Following Evans-Gariepy's book [18, p. 120], for any locally dV-integrable function f on $\mathbb{R}^n$, i.e.,

$$f \in L^1_{\text{loc}}(\mathbb{R}^n),$$

and $1 \le i \le n$, we say that

$$g_i \in L^1_{\text{loc}}(\mathbb{R}^n)$$

is the weak partial derivative of f with respect to x_i if

$$(1.1) \qquad \int_{\mathbb{R}^n} f(x) \frac{\partial \varphi(x)}{\partial x_i} \, dx = - \int_{\mathbb{R}^n} g_i(x) \varphi(x) \, dx$$

holds for all $\varphi \in C^1_c(\mathbb{R}^n)$. It is easy to check that the weak partial derivative of f with respect to x_i, if it exists, is uniquely defined for almost all $x \in \mathbb{R}^n$. We use the notation

$$\frac{\partial f}{\partial x_i} := g_i \ \forall \ i = 1, 2, \ldots, n$$

© Springer Nature Switzerland AG 2018
L. Liu et al., *Gaussian Capacity Analysis*, Lecture Notes in Mathematics 2225,
https://doi.org/10.1007/978-3-319-95040-2_1

and

$$\nabla f := \left(\frac{\partial f}{\partial x_1}, \ldots, \frac{\partial f}{\partial x_n} \right).$$

Obviously, if $f \in C^1(\mathbb{R}^n)$, then each g_i coincides with the partial derivative of f with respect to x_i.

Since $V_\gamma(\mathbb{R}^n) = 1$, we see that any

$$f \in L^p(\mathbb{G}^n) \text{ with } p \in (1, \infty]$$

belongs to $L^1(\mathbb{G}^n)$, and thus in $f \in L^1_{\text{loc}}(\mathbb{R}^n)$. Hence, it makes sense to consider the weak partial derivatives (and also ∇f) of any $f \in L^p(\mathbb{G}^n)$.

Definition 1.1.1. Let $p \in (1, \infty]$. Define the Gaussian Sobolev p-space $W^{1,p}(\mathbb{G}^n)$ to be the class of all

$$f \in L^p(\mathbb{G}^n) \text{ satisfying } \nabla f \in \left(L^p(\mathbb{G}^n) \right)^n.$$

For any $f \in W^{1,p}(\mathbb{G}^n)$, define

$$\|f\|_{W^{1,p}(\mathbb{G}^n)} := \begin{cases} \left(\|f\|^p_{L^p(\mathbb{G}^n)} + \||\nabla f|\|^p_{L^p(\mathbb{G}^n)} \right)^{\frac{1}{p}} & \text{as } p \in (1, \infty); \\ \|f\|_{L^\infty(\mathbb{G}^n)} + \||\nabla f|\|_{L^\infty(\mathbb{G}^n)} & \text{as } p = \infty. \end{cases}$$

Remark 1.1.2. By $(\star\star)$ and the Hölder inequality, it is easy to see that for

$$f \in C^1_c(\mathbb{R}^n) \text{ and } p \in (1, \infty)$$

one has

$$(1.2) \quad \|f\|_{W^{1,p}(\mathbb{G}^n)} \simeq \||\nabla f|\|_{L^p(\mathbb{G}^n)} + \|f\|_{L^1(\mathbb{G}^n)} \simeq \||\nabla f|\|_{L^p(\mathbb{G}^n)} + \left| \int_{\mathbb{R}^n} f \, dV_\gamma \right|.$$

It should be pointed out that the last sum in (1.2) coincides with the sum on the right side of $(\star\star)$.

Let $p \in (1, \infty)$. Define the little Gaussian Sobolev p-space $W^{1,p}_0(\mathbb{G}^n)$ to be the completion of all $C^1_c(\mathbb{R}^n)$-functions in

$$\left(W^{1,p}(\mathbb{G}^n), \| \cdot \|_{W^{1,p}(\mathbb{G}^n)} \right).$$

The following assertion just indicates that $W^{1,p}_0(\mathbb{G}^n)$ is nothing but $W^{1,p}(\mathbb{G}^n)$ and consequently all polynomials are dense in each Gaussian Sobolev p-space.

Proposition 1.1.3. Let $p \in (1, \infty)$. Then

$$W^{1,p}(\mathbb{G}^n) = W^{1,p}_0(\mathbb{G}^n).$$

More precisely, for any $f \in W^{1,p}(\mathbb{G}^n)$, there exists a sequence of functions

$$\{f_j\}_{j \in \mathbb{N}} \subseteq C^1_c(\mathbb{R}^n)$$

such that

(1.3)
$$\lim_{j \to \infty} \|f_j - f\|_{W^{1,p}(\mathbb{G}^n)} = 0.$$

Moreover, if K is a nonempty compact subset of $\mathbb{R}^n$ such that

$$K \subseteq \{x \in \mathbb{R}^n : f(x) \geq 1\}^{\circ},$$

then the above functions f_j enjoy that

(1.4)
$$f_j|_K \geq 1 \quad \forall \quad j \in \mathbb{N}.$$

Proof. Choose a function $\eta \in C_c^1(\mathbb{R}^n)$ satisfying

$$\begin{cases} \eta = 1 & \text{on } B(0,1); \\ \eta = 0 & \text{on } B(0,2)^c; \\ 0 \leq \eta \leq 1 & \text{on } \mathbb{R}^n; \\ |\nabla \eta| \leq 2 & \text{on } \mathbb{R}^n. \end{cases}$$

For

$$k \in \mathbb{N} \ \& \ f \in W^{1,p}(\mathbb{G}^n),$$

we observe

$$\|\eta(2^{-k}\cdot)f - f\|_{L^p(\mathbb{G}^n)} = \left(\int_{\mathbb{R}^n} |(\eta(2^{-k}x) - 1)f(x)|^p \, dV_\gamma(x) \right)^{\frac{1}{p}}$$

$$\leq \left(\int_{B(0,2^k)^c} |f(x)|^p \, dV_\gamma(x) \right)^{\frac{1}{p}},$$

which tends to 0 as $k \to \infty$ since $f \in L^p(\mathbb{G}^n)$. Meanwhile,

$$\left\| \nabla \big(\eta(2^{-k}\cdot)f \big) - \nabla f \right\|_{L^p(\mathbb{G}^n)}$$

$$= \left(\int_{\mathbb{R}^n} |2^{-k}\nabla\eta(2^{-k}x)f(x) + \eta(2^{-k}x)\nabla f(x) - \nabla f(x)|^p \, dV_\gamma(x) \right)^{\frac{1}{p}}$$

$$\leq 2^{-k+1}\|f\|_{L^p(\mathbb{G}^n)} + \left(\int_{\mathbb{R}^n} |(\eta(2^{-k}x) - 1)\nabla f(x)|^p \, dV_\gamma(x) \right)^{\frac{1}{p}}$$

$$\leq 2^{-k+1}\|f\|_{L^p(\mathbb{G}^n)} + \left(\int_{|x|\geq 2^k} |\nabla f(x)|^p \, dV_\gamma(x) \right)^{\frac{1}{p}},$$

which also tends to 0 as $k \to \infty$. Via combining the last two inequalities we deduce

$$\begin{cases} \eta(2^{-k}\cdot) \in W^{1,p}(\mathbb{G}^n); \\ \lim_{k \to \infty} \|\eta(2^{-k}\cdot)f - f\|_{W^{1,p}(\mathbb{G}^n)} = 0. \end{cases}$$

Thus, each $f \in W^{1,p}(\mathbb{G}^n)$ can be approximated by a sequence of functions in $W^{1,p}(\mathbb{G}^n)$ with compact support, and hence we can reduce the proof of (i) by assuming that

$$f \in W^{1,p}(\mathbb{G}^n)$$

has a compact support.

Without loss of generality we may assume that

$$\operatorname{supp} f \subseteq B(0, R)$$

for some fixed $R > 0$. Choose a nonnegative function $\phi \in C_c^\infty(\mathbb{R}^n)$ satisfying

$$\begin{cases} \operatorname{supp} \phi \subseteq B(0, 1); \\ \int_{\mathbb{R}^n} \phi \, dV = 1. \end{cases}$$

For any $j \in \mathbb{N}$, let

$$\phi_j(x) := 2^{jn} \phi(2^j x) \quad \forall \quad x \in \mathbb{R}^n.$$

Clearly, for all $j \in \mathbb{N}$ one has

$$\begin{cases} \operatorname{supp} \phi_j \subseteq B(0, 2^{-j}); \\ \int_{\mathbb{R}^n} \phi_j \, dV = 1. \end{cases}$$

The Hölder inequality derives

$$|\phi_j * f(x)| = \left| \int_{\mathbb{R}^n} (\phi_j(x - y))^{1 - \frac{1}{p}} (\phi_j(x - y))^{\frac{1}{p}} f(y) \, dV(y) \right|$$

$$\leq \left(\int_{\mathbb{R}^n} \phi_j(x - y) |f(y)|^p \, dV(y) \right)^{\frac{1}{p}},$$

whence

$$\int_{\mathbb{R}^n} |\phi_j * f(x)|^p \, dV_\gamma(x)$$

$$\leq \int_{\mathbb{R}^n} \int_{\mathbb{R}^n} \phi_j(x - y) |f(y)|^p \, dV(y) \, dV_\gamma(x)$$

$$= \int_{y \in B(0, R)} \left(\int_{x \in B(y, 2^{-j})} \phi_j(x - y) \gamma(y)^{-1} \gamma(x) \, dV(x) \right) |f(y)|^p \, dV_\gamma(y).$$

Upon noticing that

$$(x, y) \in B(y, 2^{-j}) \times B(0, R)$$

implies

$$\gamma(y)^{-1} \gamma(x) = e^{2^{-1}(|y|^2 - |x|^2)} \leq e^{2^{-1} R^2} =: C_R,$$

we read off

$$\int_{\mathbb{R}^n} |\phi_j * f(x)|^p \, dV_\gamma(x)$$

(1.5)
$$\leq C_R \int_{y \in B(0,R)} \left(\int_{x \in B(y, 2^{-j})} \phi_j(x - y) \, dV(x) \right) |f(y)|^p \, dV_\gamma(y)$$

$$= C_R \int_{B(0,R)} |f(y)|^p \, dV_\gamma(y) \quad \forall \ j \in \mathbb{N}.$$

Now fix $\delta > 0$. Thanks to

$$f \in L^p(\mathbb{G}^n) \ \text{ and } \ p \in (1, \infty),$$

there exists a function $g \in C_c(\mathbb{R}^n)$ such that

$$\|g - f\|_{L^p(\mathbb{G}^n)} < \delta.$$

This implies, according to (1.5), that for all $j \in \mathbb{N}$ one has

$$\|\phi_j * g - \phi_j * f\|_{L^p(\mathbb{G}^n)}^p = \|\phi_j * (g - f)\|_{L^p(\mathbb{G}^n)}^p$$

$$\leq C_R \|g - f\|_{L^p(\mathbb{G}^n)}^p$$

$$< C_R \delta^p.$$

Without loss of generality, we may assume

$$\operatorname{supp} g \subseteq B(0, R_1)$$

and proceed as in the proof of (1.5) to derive

$$\int_{\mathbb{R}^n} |\phi_j * g(x) - g(x)|^p \, dV_\gamma(x)$$

$$\leq C_{R_1} \int_{y \in B(0, R_1)} \left(\int_{x \in B(y, 2^{-j})} \phi_j(x - y)|g(y) - g(x)|^p \, dV(x) \right) dV_\gamma(y),$$

where

$$C_{R_1} := e^{2^{-1} R_1^2}.$$

Since

$$g \in C_c(\mathbb{R}^n),$$

there exists $N \in \mathbb{N}$ such that when $j > N$, we have

$$|g(y) - g(x)| < C_{R_1}^{-\frac{1}{p}} \delta \quad \text{as} \quad |x - y| < 2^{-j},$$

which implies

$$\|\phi_j * g - g\|_{L^p(\mathbb{G}^n)} \leq \delta \quad \forall \ j > N.$$

Consequently, when $j > N$ we have

$$\|\phi_j * f - f\|_{L^p(\mathbb{G}^n)}$$
$$\leq \|\phi_j * f - \phi_j * g\|_{L^p(\mathbb{G}^n)} + \|\phi_j * g - g\|_{L^p(\mathbb{G}^n)} + \|g - f\|_{L^p(\mathbb{G}^n)}$$
$$\leq (C_R^{\frac{1}{p}} + 2)\delta.$$

Letting first $j \to \infty$ and then $\delta \to 0$ yields

$$(1.6) \qquad \lim_{j \to \infty} \|\phi_j * f - f\|_{L^p(\mathbb{G}^n)} = 0.$$

Further, observing

$$\frac{\partial(\phi_j * f)(x)}{\partial x_i} = \phi_j * \frac{\partial f}{\partial x_i}(x) \quad \forall \quad (x, j, i) \in \mathbb{R}^n \times \mathbb{N} \times \{1, 2, \ldots, n\},$$

we apply (1.6) to derive

$$(1.7) \qquad \lim_{j \to \infty} \left\|\nabla(\phi_j * f) - \nabla f\right\|_{L^p(\mathbb{G}^n)} = \lim_{j \to \infty} \left\|\phi_j * \nabla f - \nabla f\right\|_{L^p(\mathbb{G}^n)} = 0.$$

Combining (1.6) and (1.7) yields that (1.3) holds for

$$f \in W^{1,p}(\mathbb{G}^n)$$

with compact support.

Finally, we show (1.4). By the above proof, we know that the following functions

$$(1.8) \qquad f_{k,j}(x) := \phi_j * \left(\eta(2^{-k}\cdot)f\right)(x) \quad \forall \quad x \in \mathbb{R}^n$$

satisfy

$$(1.9) \qquad \lim_{k \to \infty} \left(\lim_{j \to \infty} \|f_{k,j} - f\|_{W^{1,p}(\mathbb{G}^n)} \right) = 0,$$

Assume further that K is a nonempty compact subset of $\mathbb{R}^n$ and

$$K \subseteq \{x \in \mathbb{R}^n : f(x) \geq 1\}^\circ.$$

Choose $R > 0$ and $k_0 \in \mathbb{N}$ such that

$$K \subseteq B(0, 2^{-1}R) \quad \& \quad 2^{k_0} > R + 1.$$

Choose $j_0 \in \mathbb{N}$ such that

$$2^{-j_0} < \text{dist}\left(\partial K, \partial(\{x \in \mathbb{R}^n : f(x) \geq 1\}^\circ)\right).$$

Then, for any

$$(x, y) \in K \times B(x, 2^{-j_0}),$$

we observe

$$f(y) \geq 1 \quad \& \quad \eta(2^{-k_0}y) = 1.$$

Thus, when $x \in K$ we have

$$f_{k_0, j_0}(x) = \int_{\mathbb{R}^n} \phi_{j_0}(x-y)\eta(2^{-k_0}y)f(y)\,dV(y) \geq \int_{\mathbb{R}^n} \phi_{j_0}(x-y)\,dV(y) = 1.$$

This proves (1.4), thereby completing the argument for Proposition 1.1.3.   □

From Proposition 1.1.3, we deduce the following generalizations of (⋆) and Remark 1.1.2.

Corollary 1.1.4. Let $p \in (1, \infty)$. Then there exists a positive constant $C_{p,n}$ such that

(1.10) $$\left(\int_{\mathbb{R}^n} \left| f - \int_{\mathbb{R}^n} f\,dV_\gamma \right|^p dV_\gamma \right)^{\frac{1}{p}} \leq C_{p,n} \|\|\nabla f\|\|_{L^p(\mathbb{G}^n)} \quad \forall \quad f \in W^{1,p}(\mathbb{G}^n).$$

Consequently, for all

$$f \in W^{1,p}(\mathbb{G}^n)$$

one has

(1.11) $$\|f\|_{W^{1,p}(\mathbb{G}^n)} \simeq \|\|\nabla f\|\|_{L^p(\mathbb{G}^n)} + \|f\|_{L^1(\mathbb{G}^n)}$$

$$\simeq \left(\int_{\mathbb{R}^n} |\nabla f|^p \, dV_\gamma \right)^{\frac{1}{p}} + \left| \int_{\mathbb{R}^n} f\,dV_\gamma \right|.$$

Proof. Let

$$f \in W^{1,p}(\mathbb{G}^n).$$

For any $\epsilon > 0$, by Proposition 1.1.3, there exists a function

$$f_\epsilon \in C_c^1(\mathbb{R}^n)$$

such that

$$\|f - f_\epsilon\|_{W^{1,p}(\mathbb{G}^n)} < \epsilon.$$

Since

$$f_\epsilon \in C_c^1(\mathbb{R}^n),$$

it is known from (⋆) that

$$\left(\int_{\mathbb{R}^n} \left| f_\epsilon - \int_{\mathbb{R}^n} f_\epsilon \, dV_\gamma \right|^p dV_\gamma \right)^{\frac{1}{p}} \leq C_{p,n} \|\nabla f_\epsilon\|_{L^p(\mathbb{G}^n)},$$

which, together with the Minkowski inequality and the Hölder inequality, implies

$$\left(\int_{\mathbb{R}^n} \left| f - \int_{\mathbb{R}^n} f \, dV_\gamma \right|^p dV_\gamma \right)^{\frac{1}{p}}$$

$$\leq \left(\int_{\mathbb{R}^n} \left| (f - f_\epsilon) - \int_{\mathbb{R}^n} (f - f_\epsilon) \, dV_\gamma \right|^p dV_\gamma \right)^{\frac{1}{p}}$$

$$+ \left(\int_{\mathbb{R}^n} \left| f_\epsilon - \int_{\mathbb{R}^n} f_\epsilon \, dV_\gamma \right|^p dV_\gamma \right)^{\frac{1}{p}}$$

$$\leq \| f - f_\epsilon \|_{L^p(\mathbb{G}^n)} + \int_{\mathbb{R}^n} |f - f_\epsilon| \, dV_\gamma + C_{p,n} \| |\nabla f_\epsilon| \|_{L^p(\mathbb{G}^n)}$$

$$\leq 2\| f - f_\epsilon \|_{L^p(\mathbb{G}^n)} + C_{p,n} \| |\nabla f_\epsilon - \nabla f| \|_{L^p(\mathbb{G}^n)} + C_{p,n} \| |\nabla f| \|_{L^p(\mathbb{G}^n)}$$

$$\leq \epsilon(2 + C_{p,n}) + C_{p,n} \| |\nabla f| \|_{L^p(\mathbb{G}^n)}.$$

Letting $\epsilon \to 0$ in the above inequality yields (1.10). Notice that (1.11) is a consequence of (1.10) and the Hölder inequality. This finishes the proof of the corollary. □

1.2 Approximating $W^{1,p}(\mathbb{G}^n)$-Function with Cancellation

In the following proposition, we prove that if $p \in (1, \infty)$, then

$$\left\{ f \in C_c^1(\mathbb{R}^n) : \int_{\mathbb{R}^n} f \, dV_\gamma = 0 \right\}$$

is dense in

$$\left\{ f \in W^{1,p}(\mathbb{G}^n) : \int_{\mathbb{R}^n} f \, dV_\gamma = 0 \right\}.$$

Proposition 1.2.1. Let

$$\begin{cases} p \in (1, \infty); \\ f \in W^{1,p}(\mathbb{G}^n); \\ \int_{\mathbb{R}^n} f \, dV_\gamma = 0. \end{cases}$$

Then there exists a sequence of functions

$$\{ f_j \}_{j \in \mathbb{N}} \subseteq C_c^1(\mathbb{R}^n)$$

such that

(1.12)
$$\int_{\mathbb{R}^n} f_j \, dV_\gamma = 0 \quad \forall \quad j \in \mathbb{N},$$

and

(1.13)
$$\lim_{j \to \infty} \|f_j - f\|_{W^{1,p}(\mathbb{G}^n)} = 0.$$

Moreover, if K is a nonempty compact subset of $\mathbb{R}^n$ obeying

$$K \subseteq \{x \in \mathbb{R}^n : f(x) \geq 1\}^\circ,$$

then the above functions f_j enjoy

$$f_j \geq 1 \quad \text{on} \quad K \quad \forall \quad j \in \mathbb{N}.$$

Proof. Given any

$$f \in W^{1,p}(\mathbb{G}^n),$$

we shall construct a sequence of functions satisfying (1.12) and (1.13) by using $f_{k,j}$ as in (1.8).

Fix $\epsilon \in (0, \frac{1}{4})$. For any $R > 0$, we can construct an even function

$$\tau \in C_c^1(\mathbb{R})$$

such that

$$\begin{cases} 0 \leq \tau \leq 1 \\ \tau = 1 \text{ on } [0, R]; \\ \tau = 0 \text{ on } [R+1, \infty); \\ \tau \text{ is decreasing on } [R, R+1]; \\ |\tau'(t)| \leq 2 \text{ when } t \in (R, R+1). \end{cases}$$

When $R > 0$ is sufficiently large, the above construction of τ gives that

$$\int_{\mathbb{R}^n} \tau(|x|) \, dV_\gamma(x) > 2^{-2}3.$$

In this case, for some $\epsilon_0 \in (0, \frac{1}{4})$, we write

(1.14)
$$\int_{\mathbb{R}^n} \tau(|x|) \, dV_\gamma(x) = 1 - \epsilon_0.$$

In the sequel, we denote by τ_{R,ϵ_0} such a function $\tau(|\cdot|)$. By (1.9), there exist large integers k_ϵ and j_ϵ such that

(1.15)
$$\|f_{k_\epsilon, j_\epsilon} - f\|_{W^{1,p}(\mathbb{G}^n)} < \epsilon.$$

Without loss of generality, we may also choose the above k_ϵ large enough to verify

$$2^{k_\epsilon} > R + 1.$$

Choose $\lambda \in (0, 1)$ such that

$$(1 + \lambda)(1 - \epsilon_0 - \epsilon) = 1.$$

Define

$$(1.16) \qquad F_\epsilon := (1 + \lambda)\left((1 - \epsilon_0)f_{k_\epsilon, j_\epsilon} - \left(\int_{\mathbb{R}^n} f_{k_\epsilon, j_\epsilon} \, dV_\gamma\right)\tau_{R, \epsilon_0}\right).$$

By (1.14) we get

$$(1.17) \qquad \int_{\mathbb{R}^n} F_\epsilon(x) \, dV_\gamma(x) = 0.$$

By

$$\int_{\mathbb{R}^n} f \, dV_\gamma = 0,$$

the Hölder inequality and (1.15), we deduce

$$\left|\int_{\mathbb{R}^n} f_{k_\epsilon, j_\epsilon} \, dV_\gamma\right| = \left|\int_{\mathbb{R}^n} (f_{k_\epsilon, j_\epsilon} - f) \, dV_\gamma\right|$$

$$(1.18) \qquad\qquad\qquad \leq \int_{\mathbb{R}^n} |f_{k_\epsilon, j_\epsilon} - f| \, dV_\gamma$$

$$\leq \|f_{k_\epsilon, j_\epsilon} - f\|_{L^p(\mathbb{G}^n)}$$

$$< \epsilon.$$

By

$$\begin{cases} (1 + \lambda)(1 - \epsilon_0 - \epsilon) = 1; \\ \epsilon \in (0, 2^{-2}); \\ \epsilon_0 \in (0, 2^{-2}), \end{cases}$$

we observe

$$0 < (1 + \lambda)(1 - \epsilon_0) - 1$$
$$= \epsilon(1 + \lambda)$$
$$= \frac{\epsilon}{1 - \epsilon_0 - \epsilon}$$
$$< 2\epsilon.$$

From this, (1.16) and (1.18), it follows that if

$$\Lambda = (1 + \lambda)(1 - \epsilon_0),$$

then

$$|F_\epsilon - f| \leq \Lambda \left|f_{k_\epsilon, j_\epsilon} - f\right| + ((1 + \lambda)(1 - \epsilon_0) - 1)|f| + (1 + \lambda)\tau_{R, \epsilon_0} \left|\int_{\mathbb{R}^n} f_{k_\epsilon, j_\epsilon} \, dV_\gamma\right|$$

$$\leq \Lambda \left|f_{k_\epsilon, j_\epsilon} - f\right| + 2\epsilon|f| + 2\epsilon,$$

which, together with (1.15), further gives

(1.19)
$$\|F_\epsilon - f\|_{L^p(\mathbb{G}^n)} \leq \Lambda\|f_{k_\epsilon, j_\epsilon} - f\|_{L^p(\mathbb{G}^n)} + 2\epsilon\|f\|_{L^p(\mathbb{G}^n)} + 2\epsilon$$

$$\leq 2\epsilon\big(2 + \|f\|_{L^p(\mathbb{G}^n)}\big).$$

To estimate $\|\nabla F_\epsilon - \nabla f\|_{L^p(\mathbb{G}^n)}$, we have

$$|\nabla F_\epsilon - \nabla f| \leq \Lambda\left(\left|\nabla f_{k_\epsilon, j_\epsilon} - \nabla f\right| + \left|1 - \Lambda^{-1}\right||\nabla f| + \frac{|\nabla \tau_{R, \epsilon_0}|}{1 - \epsilon_0}\left|\int_{\mathbb{R}^n} f_{k_\epsilon, j_\epsilon} \, dV_\gamma\right|\right)$$

$$\leq \Lambda\left|\nabla f_{k_\epsilon, j_\epsilon} - \nabla f\right| + 2\epsilon|\nabla f| + 4\epsilon,$$

where we have used

$$\left\|\,|\nabla \tau|\,\right\|_{L^\infty(\mathbb{R}^n)} \leq 2,$$

(1.16) and (1.18). Consequently,

(1.20)
$$\left\|\,|\nabla F_\epsilon - \nabla f|\,\right\|_{L^p(\mathbb{G}^n)}$$

$$\leq \Lambda\left\|\,|\nabla f_{k_\epsilon, j_\epsilon} - \nabla f|\,\right\|_{L^p(\mathbb{G}^n)} + 2\epsilon\left\|\,|\nabla f|\,\right\|_{L^p(\mathbb{G}^n)} + 4\epsilon$$

$$\leq 2\left(3 + \left\|\,|\nabla f|\,\right\|_{L^p(\mathbb{G}^n)}\right)\epsilon.$$

Summarizing the above arguments, we derive that for any $\epsilon \in (0, 2^{-2})$, there exists a function F_ϵ satisfying (1.17), (1.19), and (1.20). This finishes the first part of the proposition.

Assume further that K is a nonempty compact subset of $\mathbb{R}^n$ and

$$K \subseteq \{x \in \mathbb{R}^n : f(x) \geq 1\}^\circ.$$

For this case, we choose R such that

$$K \subseteq B(0, 2^{-1}R).$$

We may as well assume that the above chosen j_ϵ also satisfies

$$2^{-j_\epsilon} < \text{dist}\left(\partial K, \partial(\{x \in \mathbb{R}^n : f(x) \geq 1\}^\circ)\right),$$

which implies

(1.21)
$$f(y) \geq 1 \quad \forall \, (x, y) \in K \times B(x, 2^{-j_\epsilon}).$$

Since k_ϵ was chosen to satisfy

$$2^{k_\epsilon} > R + 1,$$

by the definition of η in the proof of Proposition 1.1.3 and especially the fact that

$$\eta(x) = 1 \quad \forall \quad |x| < 1,$$

we observe

$$2^{-k_\epsilon}|y| \leq 2^{-k_\epsilon}(2^{-1}R + 1) < 1 \quad \forall \quad (x, y) \in K \times B(x, 2^{-j_\epsilon}),$$

so that

(1.22) $$\eta(2^{-k_\epsilon}y) = 1 \quad \forall \quad (x, y) \in K \times B(x, 2^{-j_\epsilon}).$$

Based on these, using (1.9), $\tau_\epsilon = 1$ on K, (1.18), (1.21), and (1.22), we obtain that if $x \in K$, then

$$F_\epsilon(x) = (1 + \lambda)\left((1 - \epsilon_0)f_{k_\epsilon, j_\epsilon}(x) - \tau_{R,\epsilon_0}(x)\int_{\mathbb{R}^n} f_{k_\epsilon, j_\epsilon}\,dV_\gamma\right)$$

$$\geq (1 + \lambda)\left((1 - \epsilon_0)\int_{\mathbb{R}^n} \phi_{j_\epsilon}(x - y)\eta(2^{-k_\epsilon}y)f(y)\,dV(y) - \epsilon\right)$$

$$\geq (1 + \lambda)\left((1 - \epsilon_0)\int_{\mathbb{R}^n} \phi_{j_\epsilon}(x - y)\,dV(y) - \epsilon\right)$$

$$= (1 + \lambda)(1 - \epsilon_0 - \epsilon)$$

$$= 1.$$

Thus, we complete the proof of Proposition 1.2.1. □

1.3 Compactness for $W^{1,p}(\mathbb{G}^n)$

For any two functions f and g, define their Gaussian inner product:

$$\langle f, g \rangle_\gamma = \int_{\mathbb{R}^n} f\,g\,dV_\gamma.$$

For any $p \in (1, \infty]$, the symbol $L^p(\mathbb{G}^n; \mathbb{R}^n)$ represents the space of vector-valued functions

$$f = (f_1, \ldots, f_n)$$

enjoying

$$f_i \in L^p(\mathbb{G}^n) \quad \forall \quad i \in \{1, 2, \ldots, n\}.$$

Proposition 1.3.1. Let $p \in (1, \infty)$. Assume that the sequence $\{f_k\}_{k \in \mathbb{N}}$ satisfies

(1.23) $$\sup_{k \in \mathbb{N}} \|f_k\|_{W^{1,p}(\mathbb{G}^n)} < \infty.$$

Then there exist a subsequence

$$\{f_{k_i}\}_{i \in \mathbb{N}} \subseteq \{f_k\}_{k \in \mathbb{N}}$$

and a function

$$f \in W^{1,p}(\mathbb{G}^n)$$

such that

$$\{(f_{k_i}, \nabla f_{k_i})\}_{i \in \mathbb{N}}$$

converges to $(f, \nabla f)$ weakly in

$$L^p(\mathbb{G}^n) \times L^p(\mathbb{G}^n; \mathbb{R}^n),$$

that is,

(1.24) $$\lim_{i \to \infty} \langle f_{k_i}, \phi \rangle_\gamma = \langle f, \phi \rangle_\gamma \quad \forall \, \phi \in L^{p'}(\mathbb{G}^n)$$

and

(1.25) $$\lim_{i \to \infty} \langle \nabla f_{k_i}, \Phi \rangle_\gamma = \langle \nabla f, \Phi \rangle_\gamma \quad \forall \, \Phi \in L^{p'}(\mathbb{G}^n; \mathbb{R}^n).$$

where p' is the conjugate index of p. Moreover,

(1.26) $$\lim_{i \to \infty} \int_{\mathbb{R}^n} f_{k_i} \, dV_\gamma = \int_{\mathbb{R}^n} f \, dV_\gamma.$$

Proof. (1.23) induces a subsequence $\{f_{k_i}\}_{i \in \mathbb{N}}$ such that

$$\{(f_{k_i}, \nabla f_{k_i})\}_{i \in \mathbb{N}}$$

tends to some (f, F) weakly in

$$L^p(\mathbb{G}^n) \times L^p(\mathbb{G}^n; \mathbb{R}^n).$$

This indicates that not only (1.24) holds, but also (1.25) holds with ∇f there replaced by F.

To get (1.25) fully, we need to verify

$$F = \nabla f.$$

Toward this end, recall Mazur's Theorem in [57, p. 120, Theorem 2] - if $\{x_j\}$ in a normed linear space $(\mathsf{X}, \|\cdot\|)$ converges weakly to x_∞, then for any $\epsilon > 0$ there is a convex combination of $\{x_i\}$:

$$\begin{cases} \sum_{i=1}^{j} \alpha_i x_i; \\ 0 \le \alpha_i \le 1; \\ \sum_{i=1}^{j} \alpha_i = 1, \end{cases}$$

such that

$$\left\| x_\infty - \sum_{i=1}^{j} \alpha_i x_i \right\| < \epsilon.$$

Then we obtain a convex combination

$$\sum_{i=1}^{m} \lambda_{m,i}(f_{k_i}, \nabla f_{k_i})$$

converging to (f, F) strongly in

$$L^p(\mathbb{G}^n) \times L^p(\mathbb{G}^n; \mathbb{R}^n),$$

where

$$\lambda_{m,i} \in (0, 1] \quad \& \quad \sum_{i=1}^{m} \lambda_{m,i} = 1.$$

In particular, we have

$$\lim_{m \to \infty} \sum_{i=1}^{m} \lambda_{m,i} f_{k_i} = f \quad \text{in} \quad L^p(\mathbb{G}^n)$$

and

$$\lim_{m \to \infty} \sum_{i=1}^{m} \lambda_{m,i} \nabla f_{k_i} = F \quad \text{in} \quad L^p(\mathbb{G}^n; \mathbb{R}^n),$$

thereby getting

$$F = \nabla f \quad \text{a.e. on } \mathbb{R}^n.$$

This completely proves (1.25).

Finally, we take $\phi = 1$ in (1.24) and then obtain (1.26). $\square$

1.4 Poincaré or Log-Sobolev Inequality for $W^{1,2}(\mathbb{G}^n)$

As shown in [25], the space $W^{1,2}(\mathbb{G}^n)$ deserves special treatment. Due to

$$\nabla \gamma(x) = -x \cdot \gamma(x) \quad \forall \quad x \in \mathbb{R}^n,$$

we can utilize the chain rule to obtain that if $f, g \in C_c^1(\mathbb{R}^n)$ and $i \in \{1, \dots, n\}$, then

$$\int_{\mathbb{R}^n} \left(\frac{\partial f(x)}{\partial x_i} \right) g(x) \, dV_\gamma(x) = -\int_{\mathbb{R}^n} f(x) \frac{\partial (g(x)\gamma(x))}{\partial x_i} \, dx$$

$$= -\int_{\mathbb{R}^n} f(x) \left(\frac{\partial g(x)}{\partial x_i} - x_i g(x) \right) dV_\gamma(x).$$

Upon introducing

$$\mathcal{L}g(x) := -\nabla g(x) + xg(x),$$

we see that $\mathcal{L}$ exists as the adjoint operator of ∇ and enjoys the integration-by-parts formula of Malliavin calculus

(1.27) $$\int_{\mathbb{R}^n} g\nabla f \, dV_\gamma = \int_{\mathbb{R}^n} f\mathcal{L}g \, dV_\gamma \quad \forall \, (f,g) \in W^{1,2}(\mathbb{G}^n) \times C_c^1(\mathbb{R}^n)$$

thanks to Proposition 1.1.3.

Moreover, if

$$\mathcal{L}f(x) := (\nabla \cdot \nabla)f(x) - x \cdot \nabla f(x) = \Delta f(x) - x \cdot \nabla f(x) \quad \forall \, (f,g) \in C_c^2(\mathbb{R}^n),$$

then

$$\int_{\mathbb{R}^n} g\mathcal{L}f \, dV_\gamma = \int_{\mathbb{R}^n} g\Delta f \, dV_\gamma - \int_{\mathbb{R}^n} (g(x)x \cdot \nabla f(x)) \, dV_\gamma(x)$$

$$= -\int_{\mathbb{R}^n} \nabla(g(x)\gamma(x)) \cdot \nabla f(x) \, dx - \int_{\mathbb{R}^n} g(x)x \cdot \nabla f(x) \, dV_\gamma(x)$$

$$= -\int_{\mathbb{R}^n} \nabla g \cdot \nabla f \, dV_\gamma$$

$$= \int_{\mathbb{R}^n} f\mathcal{L}g \, dV_\gamma,$$

and hence $\mathcal{L}$ is symmetric in $L^2(\mathbb{G}^n)$.

Proposition 1.4.1. For a bounded and continuous function f on $\mathbb{R}^n$, denoted by

$$f \in C_b^0(\mathbb{R}^n),$$

let

$$P_t f(x) = \int_{\mathbb{R}^n} f\left(e^{-t}x + \sqrt{1 - e^{-2t}}y\right) dV_\gamma(y) \quad \forall \, (x,t) \in \mathbb{R}^n \times [0,\infty).$$

Then the Ornstein-Uhlenbeck semigroup $(P_t)_{t\in[0,\infty)}$ satisfies the following properties.

(i) $P_0 f = f \ \forall \ f \in C_b^0(\mathbb{R}^n)$.

(ii) $t \mapsto P_t f$ is continuous from $[0,\infty)$ to $L^2(\mathbb{G}^n)$.

(iii) $P_t \circ P_s = P_{s+t} \ \forall \ s,t \in [0,\infty)$.

(iv) $P_t 1 = 1 \ \& \ P_t f \geq 0 \ \forall \ t \geq 0 \ \& \ 0 \leq f \in C_b^0(\mathbb{R}^n)$.

(v) $\|P_t f\|_{L^\infty(\mathbb{R}^n)} \leq \|f\|_{L^\infty(\mathbb{R}^n)} \ \forall \ f \in C_b^0(\mathbb{R}^n)$.

(vi) $\partial_t P_t f = \mathcal{L}(P_t f) = P_t(\mathcal{L}f) \ \forall \ (t,f) \in [0,\infty) \times C_c^2(\mathbb{R}^n)$.

(vii) $\lim_{t\to\infty} P_t f(x) = \int_{\mathbb{R}^n} f\, dV_\gamma \ \forall\ (x, f) \in \mathbb{R}^n \times C_b^0(\mathbb{R}^n)$.

(viii) $\int_{\mathbb{R}^n} \mathcal{L}f\, dV_\gamma = 0 \ \forall\ f \in C_c^2(\mathbb{R}^n)$.

(ix) $|\nabla P_t f(x)| \le e^{-t} P_t |\nabla f|(x) \ \forall\ (t, f, x) \in [0, \infty) \times C_c^1(\mathbb{R}^n) \times \mathbb{R}^n$.

Proof. It suffices to verify (vi) and (ix). For (vi) we use

$$f \in C_c^1(\mathbb{R}^n),$$

the definition of $P_t f$ and (1.27) to compute

$$\partial_t P_t f(x) = \int_{\mathbb{R}^n} \left(\frac{e^{-2t} y}{\sqrt{1 - e^{-2t}}} - e^{-t} x \right) \cdot \nabla f\left(e^{-t}x + \sqrt{1 - e^{-2t}} y\right) dV_\gamma(y)$$

$$= \left(\frac{e^{-2t}}{\sqrt{1 - e^{-2t}}} \right) \int_{\mathbb{R}^n} y \cdot \nabla f\left(e^{-t}x + \sqrt{1 - e^{-2t}} y\right) dV_\gamma(y)$$

$$- e^{-t} x \cdot \int_{\mathbb{R}^n} \nabla f\left(e^{-t}x + \sqrt{1 - e^{-2t}} y\right) dV_\gamma(y)$$

$$= \Delta P_t f(x) - x \cdot \nabla P_t f(x)$$

$$= \mathcal{L}(P_t f)(x).$$

For (ix) we calculate

$$\nabla P_t f(x) = e^{-t} \int_{\mathbb{R}^n} \nabla f\left(e^{-t}x + \sqrt{1 - e^{-2t}} y\right) dV_\gamma(y),$$

thereby reaching the pointwise inequality. $\qquad\square$

Theorem 1.4.2. For

$$f \in C_c^1(\mathbb{R}^n)$$

let

$$\mathrm{Var}_\gamma(f) := \int_{\mathbb{R}^n} \left(f - \int_{\mathbb{R}^n} f\, dV_\gamma \right)^2 dV_\gamma$$

and

$$\mathrm{Ent}_\gamma(f^2) := \int_{\mathbb{R}^n} f^2 \left(\ln \frac{f^2}{\int_{\mathbb{R}^n} f^2\, dV_\gamma} \right) dV_\gamma$$

be the Gauss variance of f and the Gauss entropy of f^2 respectively.

(i) The Poincaré inequality

$$\mathrm{Var}_\gamma(f) \le \int_{\mathbb{R}^n} |\nabla f|^2\, dV_\gamma,$$

equivalently,

$$\mathrm{Var}_\gamma(P_t f) \le e^{-2t} \mathrm{Var}_\gamma(f) \ \forall\ t \in [0, \infty),$$

holds. Moreover, both inequalities are optimal and extremal functions are determined by $\nabla f = c$ for some constant $c \in \mathbb{R}^n$.

(ii) The logarithmic Sobolev inequality

$$\mathrm{Ent}_\gamma(f^2) \leq 2 \int_{\mathbb{R}^n} |\nabla f|^2 \, dV_\gamma,$$

equivalently,

$$\mathrm{Ent}_\gamma\left(P_t(f^2)\right) \leq e^{2t} \mathrm{Ent}_\gamma(f^2) \quad \forall \quad t \in [0, \infty),$$

holds. Moreover, both inequalities are optimal and extremal functions are determined by

$$\nabla f = cf \text{ for some constant } c \in \mathbb{R}^n.$$

Proof. (i) According to Proposition 1.4.1 and the Cauchy-Schwarz inequality, we obtain

$$\begin{aligned}
\mathrm{Var}_\gamma(f) &= -\int_0^\infty \partial_t \left(\int_{\mathbb{R}^n} (P_t f)^2 \, dV_\gamma \right) dt \\
&= -2 \int_0^\infty \int_{\mathbb{R}^n} \mathcal{L} P_t f P_t f \, dV_\gamma dt \\
&= 2 \int_0^\infty \int_{\mathbb{R}^n} |\nabla P_t f|^2 \, dV_\gamma dt \\
&\leq 2 \int_0^\infty \int_{\mathbb{R}^n} e^{-2t} \left(P_t |\nabla f|^2 \right) dV_\gamma dt \\
&\leq 2 \int_0^\infty \int_{\mathbb{R}^n} e^{-2t} P_t \left(|\nabla f|^2 \right) dV_\gamma dt \\
&= 2 \int_0^\infty \int_{\mathbb{R}^n} e^{-2t} |\nabla f|^2 \, dV_\gamma dt \\
&= \int_{\mathbb{R}^n} |\nabla f|^2 \, dV_\gamma.
\end{aligned}$$

This, along with

$$\partial_t \mathrm{Var}_\gamma(P_t f) = -2 \int_{\mathbb{R}^n} |\nabla P_t f|^2 \, dV_\gamma$$

and the Grönwall lemma, implies

$$\mathrm{Var}_\gamma(P_t f) \leq e^{-2t} \mathrm{Var}_\gamma(f) \quad \forall \quad t \in [0, \infty).$$

Conversely, if the last inequality holds, then its derivation at time $t = 0$ derives

$$\mathrm{Var}_\gamma(f) \leq \int_{\mathbb{R}^n} |\nabla f|^2 \, dV_\gamma.$$

While checking all stages of the argument for this last inequality, we can see that smooth functions obeying $\nabla f = c$ are the unique function ensuring the equalities in the two inequalities of (i).

(ii) For simplicity, let
$$g = f^2 \in C_c^1(\mathbb{R}).$$

Just like the argument for (i) we have

$$
\begin{aligned}
\mathrm{Ent}_\gamma(g) &= -\int_0^\infty \partial_t\left(\int_{\mathbb{R}^n} (P_t g)\ln(P_t g)\,dV_\gamma\right)dt \\
&= -\int_0^\infty \int_{\mathbb{R}^n} \mathcal{L}P_t g\ln(P_t g)\,dV_\gamma dt \\
&= \int_0^\infty \int_{\mathbb{R}^n} (\nabla P_t g)\cdot(\nabla\ln(P_t g))\,dV_\gamma dt \\
&= \int_0^\infty \int_{\mathbb{R}^n} \frac{|\nabla P_t g|^2}{P_t g}\,dV_\gamma dt \\
&\le \int_0^\infty \int_{\mathbb{R}^n} e^{-2t}\frac{(P_t|\nabla g|)^2}{P_t g}\,dV_\gamma dt.
\end{aligned}
$$

Now, the Cauchy-Schwarz inequality or the convexity of the map

$$(0,\infty)\times(0,\infty) \ni (s,t) \mapsto s^2 t^{-1},$$

derives

$$\frac{(P_t|\nabla g|)^2}{P_t g} \le P_t\left(\frac{|\nabla g|^2}{g}\right),$$

whence

$$\mathrm{Ent}_\gamma(g) \le \int_0^\infty \int_{\mathbb{R}^n} e^{-2t} P_t\left(\frac{|\nabla g|^2}{g}\right)dV_\gamma dt = 2^{-1}\int_{\mathbb{R}^n}\frac{|\nabla g|^2}{g}\,dV_\gamma.$$

In fact, this last inequality amounts to

$$\mathrm{Ent}_\gamma(P_t g) \le e^{2t}\,\mathrm{Ent}_\gamma(g) \quad \forall\ \ t\in[0,\infty),$$

due to the following variation formula

$$\partial_t\,\mathrm{Ent}_\gamma(P_t g) = -\int_{\mathbb{R}^n}\frac{|\nabla P_t g|^2}{P_t g}\,dv_\gamma.$$

Similarly, we can obtain the extremal functions for (ii) via examining the equalities for (i), i.e., the functions satisfying

$$\nabla(f^2) = cf^2 \ \text{ for a constant } \ c\in\mathbb{R}^n.$$

$\square$

Chapter 2
Gaussian Campanato (p, κ)-Class

In this chapter we are motivated by the left side of $(\star)$ to investigate the Campanato (p, κ)-class on $\mathbb{G}^n$ and its relationship with the Morrey space, John-Nirenberg space, and Lipschitz space on $\mathbb{G}^n$.

2.1 Location of $C^{p,\kappa}(\mathbb{G}^n)$

In what follows, define

$$m(x) := \min\{1, |x|^{-1}\} \quad \forall \quad x \in \mathbb{R}^n.$$

For any $a \in (0, \infty)$ denote by $\mathcal{B}_a$ the set of all balls B in $\mathbb{R}^n$ such that

$$r_B \le a\, m(c_B),$$

where c_B and r_B denote the center and the radius of B respectively. Balls in $\mathcal{B}_a$ are called admissible balls with scale a. If

$$(B, x) \in \mathcal{B}_a \times B,$$

then

(2.1) $$(a + 1)^{-1} m(x) \le m(c_B) \le (a + 1)m(x)$$

and

(2.2) $$e^{-\frac{a^2+2a}{2}} \le e^{\frac{|c_B|^2-|x|^2}{2}} \le e^a;$$

see [40, Proposition 2.1]. Consequently, for all $B \in \mathcal{B}_a$,

(2.3) $$e^{-\frac{a^2+2a}{2}} \lesssim \frac{V_\gamma(B)}{e^{-\frac{|c_B|^2}{2}} r_B^n} \lesssim e^a.$$

This inequality implies that dV_γ is doubling on the admissible class $\mathcal{B}_a$, that is, for all balls $B \in \mathcal{B}_a$ one has

(2.4) $$V_\gamma(2B) \lesssim V_\gamma(B) \quad \text{(locally doubling property)}.$$

© Springer Nature Switzerland AG 2018
L. Liu et al., *Gaussian Capacity Analysis*, Lecture Notes in Mathematics 2225,
https://doi.org/10.1007/978-3-319-95040-2_2

A locally integrable function f is said to be of bounded mean oscillation on $\mathbb{G}^n$, denoted by $\mathrm{BMO}(\mathbb{G}^n)$, provided that

$$\|f\|_{\mathrm{BMO}(\mathbb{G}^n)} := \sup_{B \in \mathcal{B}_a} \frac{1}{V_\gamma(B)} \int_B |f(x) - f_{B,\gamma}|\, dV_\gamma(x) < \infty.$$

In the above and below,

(2.5)
$$f_{B,\gamma} := \frac{1}{V_\gamma(B)} \int_B f\, dV_\gamma.$$

It was proved in [40] that, for all $p \in (1, \infty)$,

$$f \in L^1(\mathbb{G}^n) \cap \mathrm{BMO}(\mathbb{G}^n)$$

$$\Leftrightarrow \|f\|_{L^1(\mathbb{G}^n)} + \sup_{B \in \mathcal{B}_a} \left(\frac{1}{V_\gamma(B)} \int_B |f(x) - f_{B,\gamma}|^p\, dV_\gamma(x) \right)^{\frac{1}{p}} < \infty$$

with implicit constants depending only on a, p, and n.

Definition 2.1.1. Let $a \in (0, \infty)$, $p \in [1, \infty)$, and $\kappa \in (-\infty, 1]$.

(i) Any locally integrable function f is said to be in the Gaussian Campanato (p, κ)-class $C_{\mathcal{B}_a}^{p,\kappa}(\mathbb{G}^n)$ provided

$$\|f\|_{C_{\mathcal{B}_a}^{p,\kappa}(\mathbb{G}^n)} := \sup_{B \in \mathcal{B}_a} \left(\frac{1}{V_\gamma(B)^{1-\kappa}} \int_B |f(x) - f_{B,\gamma}|^p\, dV_\gamma(x) \right)^{\frac{1}{p}} < \infty,$$

where $f_{B,\gamma}$ is as in (2.5). Moreover,

$$\begin{cases} \mathcal{L}_{\mathcal{B}_a}^{p,\kappa}(\mathbb{G}^n) := L^1(\mathbb{G}^n) \cap C_{\mathcal{B}_a}^{p,\kappa}(\mathbb{G}^n); \\ \|f\|_{\mathcal{L}_{\mathcal{B}_a}^{p,\kappa}(\mathbb{G}^n)} := \|f\|_{L^1(\mathbb{G}^n)} + \|f\|_{C_{\mathcal{B}_a}^{p,\kappa}(\mathbb{G}^n)}. \end{cases}$$

When $a = 1$, we write

$$C_{\mathcal{B}_a}^{p,\kappa}(\mathbb{G}^n) \ \& \ \mathcal{L}_{\mathcal{B}_a}^{p,\kappa}(\mathbb{G}^n) \text{ as } C^{p,\kappa}(\mathbb{G}^n) \ \& \ \mathcal{L}^{p,\kappa}(\mathbb{G}^n).$$

(ii) $\mathcal{Q}_a$ consists of all admissible cubes Q with sides parallel to the axes, the center c_Q, and the side-length

$$\ell_Q \le a\, m(c_Q).$$

Then

$$\begin{cases} C_{\mathcal{Q}_a}^{p,\kappa}(\mathbb{G}^n) \ \& \ \mathcal{L}_{\mathcal{Q}_a}^{p,\kappa}(\mathbb{G}^n); \\ \| \cdot \|_{C_{\mathcal{Q}_a}^{p,\kappa}(\mathbb{G}^n)} \ \& \ \| \cdot \|_{\mathcal{L}_{\mathcal{Q}_a}^{p,\kappa}(\mathbb{G}^n)}, \end{cases}$$

are defined via replacing $B \in \mathcal{B}_a$ with $Q \in \mathcal{Q}_a$ in (i). Simply, we write

$$\begin{cases} C_{\mathcal{B}_a}^{p,\kappa}(\mathbb{G}^n) \text{ or } C_{\mathcal{Q}_a}^{p,\kappa}(\mathbb{G}^n) \text{ as } C^{p,\kappa}(\mathbb{G}^n); \\ \mathcal{L}_{\mathcal{B}_a}^{p,\kappa}(\mathbb{G}^n) \text{ or } \mathcal{L}_{\mathcal{Q}_a}^{p,\kappa}(\mathbb{G}^n) \text{ as } \mathcal{L}^{p,\kappa}(\mathbb{G}^n), \end{cases}$$

if there exists no confusion.

Lemma 2.1.2. For any cube $Q \subseteq \mathbb{R}^n$ and $p \in [1, \infty]$, denote by $L_0^p(Q; \mathbb{G}^n)$ the family of all functions

$$\phi \in L^p(\mathbb{G}^n) \text{ satisfying } \operatorname{supp} \phi \subseteq Q \ \& \ \int_Q \phi \, dV_\gamma = 0.$$

(i) If

$$\begin{cases} 0 < b < a < \infty; \\ p \in [1, \infty); \\ \kappa \in (-\infty, 1], \end{cases}$$

then

$$\|f\|_{C_{\mathcal{B}_a}^{p,\kappa}(\mathbb{G}^n)} \simeq \|f\|_{C_{Q_a}^{p,\kappa}(\mathbb{G}^n)} \simeq \|f\|_{C_{Q_b}^{p,\kappa}(\mathbb{G}^n)} \simeq \|f\|_{C_{\mathcal{B}_b}^{p,\kappa}(\mathbb{G}^n)}$$

with implicit constants depending only on n, a, b, p, and κ. Consequently,

$$\mathcal{L}_{\mathcal{B}_a}^{p,\kappa}(\mathbb{G}^n) = \mathcal{L}_{Q_a}^{p,\kappa}(\mathbb{G}^n) = \mathcal{L}_{Q_b}^{p,\kappa}(\mathbb{G}^n) = \mathcal{L}_{\mathcal{B}_b}^{p,\kappa}(\mathbb{G}^n).$$

(ii) If

$$\begin{cases} p \in [1, \infty); \\ \kappa \in (-\infty, 0]; \\ f \in C^{p,\kappa}(\mathbb{G}^n), \end{cases}$$

then for all

$$(B, B') \in \mathcal{B}_1 \times \mathcal{B}_1 \text{ with } c_B = c_{B'} \ \& \ r_B < r_{B'},$$

it holds that

$$(2.6) \quad |f_{B,\gamma} - f_{B',\gamma}| \lesssim \begin{cases} \left(1 + \ln \frac{r_{B'}}{r_B}\right) \|f\|_{C^{p,\kappa}(\mathbb{G}^n)} & \text{as } \kappa = 0; \\ \left(\frac{r_B}{r_{B'}}\right)^{\frac{n\kappa}{p}} \dfrac{\|f\|_{C^{p,\kappa}(\mathbb{G}^n)}}{V_\gamma(B)^{\frac{\kappa}{p}}} & \text{as } \kappa \in (-\infty, 0), \end{cases}$$

with implicit constants depending only on p, κ, and n.

Proof. (i) By a reexamination of the proof of [40, Lemma 2.3], we can get a nonnegative integer N (depending only on n, p, a, and b) such that for any cube $Q \in Q_a$ and each function

$$\phi \in L_0^{p'}(Q; \mathbb{G}^n),$$

there exist at most N subcubes $\{Q_1, \ldots, Q_N\}$ in the admissible class Q_b and N functions

$$\{\phi_1, \ldots, \phi_N\} \in L_0^{p'}(Q; \mathbb{G}^n)$$

such that

$$\begin{cases} \operatorname{supp} \phi_j \subseteq Q_j \quad \forall \ j \in \{1, \ldots, N\}; \\ \phi = \sum_{j=1}^N \phi_j; \\ \|\phi_j\|_{L^p(\mathbb{G}^n)} \lesssim \|\phi\|_{L^p(\mathbb{G}^n)}; \\ V_\gamma(Q_j) \simeq V_\gamma(Q) \quad \forall \ j \in \{1, \ldots, N\}. \end{cases}$$

The rest of the argument is completely parallel to that for [40, Proposition 2.4], so the details are omitted.

(ii) We show (2.6) by considering the following two cases:

$$r_B \geq 2^{-1} r_{B'} \quad \& \quad r_B < 2^{-1} r_{B'}.$$

Case $r_B \geq 2^{-1} r_{B'}$. In this case, we have

$$B \subseteq B' \subseteq 2B,$$

and hence

$$V_\gamma(B) \simeq \gamma(B')$$

by (2.4). From this and Hölder's inequality, we deduce

$$(2.7) \quad |f_{B,\gamma} - f_{B',\gamma}| \leq \frac{1}{V_\gamma(B)} \int_B |f(x) - f_{B',\gamma}| \, dV_\gamma(x) \lesssim \frac{\|f\|_{C^{p,\kappa}(\mathbb{G}^n)}}{V_\gamma(B)^{\frac{\kappa}{p}}},$$

thereby reaching (2.6).

Case $r_B < 2^{-1} r_{B'}$. Under this situation we choose $j_0 \geq 1$ such that

$$2^{j_0} r_B \leq r_{B'} < 2^{j_0+1} r_B.$$

Via setting

$$B_j := B(c_B, 2^j r_B) \quad \forall \quad j \in \{1, \dots, j_0\},$$

we have

$$|f_{B,\gamma} - f_{B',\gamma}| \leq \sum_{j=0}^{j_0-1} |f_{B_j,\gamma} - f_{B_{j+1},\gamma}| + |f_{B_{j_0},\gamma} - f_{B',\gamma}|.$$

For any

$$j \in \{1, \dots, j_0 - 1\},$$

we use

$$r_{B_j} = 2^{-1} r_{B_{j+1}}$$

and (2.7) to conclude

$$(2.8) \qquad |f_{B_j,\gamma} - f_{B_{j+1},\gamma}| \lesssim \frac{\|f\|_{C^{p,\kappa}(\mathbb{G}^n)}}{V_\gamma(B_j)^{\frac{\kappa}{p}}}.$$

Similarly, $|f_{B_{j_0}} - f_{B',\gamma}|$ has the same upper bound as in (2.8). Accordingly, via summing the inequality (2.8), we know

$$|f_{B,\gamma} - f_{B',\gamma}| \lesssim \sum_{j=0}^{j_0} \frac{\|f\|_{C^{p,\kappa}(\mathbb{G}^n)}}{V_\gamma(B_j)^{\frac{\kappa}{p}}} \lesssim \frac{\|f\|_{C^{p,\kappa}(\mathbb{G}^n)}}{V_\gamma(B)^{\frac{\kappa}{p}}} \sum_{j=0}^{j_0} 2^{-\frac{jn\kappa}{p}},$$

where we have used the fact that

$$\gamma(B_j) \simeq 2^{jn} V_\gamma(B).$$

Then, applying

$$2^{j_0} \simeq \frac{r_{B'}}{r_B},$$

we see that (2.6) holds for the case $r_B < 2^{-1} r_{B'}$.   □

Using Lemma 2.1.2, we can locate each Gaussian Campanato class.

Proposition 2.1.3. Let $p \in [1, \infty)$.

(i) If $\kappa \in (0, 1)$, then

$$\mathcal{L}^{p,\kappa}(\mathbb{G}^n) = \mathcal{M}^{p,\kappa}(\mathbb{G}^n),$$

the Morrey space of all functions f obeying

$$\|f\|_{\mathcal{M}^{p,\kappa}(\mathbb{G}^n)} := \|f\|_{L^1(\mathbb{G}^n)} + \sup_{B \in \mathcal{B}_1} \left(V_\gamma(B)^{\kappa-1} \int_B |f|^p \, dV_\gamma \right)^{\frac{1}{p}} < \infty.$$

(ii) If $\kappa = 0$, then

$$C^{p,\kappa}(\mathbb{G}^n) = \mathrm{BMO}(\mathbb{G}^n).$$

(iii) If $\kappa \in [-\frac{p}{n}, 0)$, then

$$C^{p,\kappa}(\mathbb{G}^n) = \mathrm{Lip}_{-\frac{\kappa}{p}}(\mathbb{G}^n),$$

the Lipschitz space of all locally integrable functions f obeying

$$\|f\|_{\mathrm{Lip}_{-\frac{\kappa}{p}}(\mathbb{G}^n)} := \sup_{B \in \mathcal{B}_1 \ \& \ x,y \in B} V_\gamma(B)^{\frac{\kappa}{p}} |f(x) - f(y)| < \infty.$$

(iv) If $\kappa \in (-\infty, -\frac{p}{n})$, then $C^{p,\kappa}(\mathbb{G}^n)$ consists of only functions which are constant almost everywhere.

Proof. (i) Clearly, if

$$f \in \mathcal{M}^{p,\kappa}(\mathbb{G}^n),$$

then an application of the Hölder inequality gives

$$\|f\|_{\mathcal{L}^{p,\kappa}(\mathbb{G}^n)} \lesssim \|f\|_{\mathcal{M}^{p,\kappa}(\mathbb{G}^n)}.$$

and so

$$\mathcal{M}^{p,\kappa}(\mathbb{G}^n) \subseteq \mathcal{L}^{p,\kappa}(\mathbb{G}^n).$$

Conversely, we need only to show that if

$$(f, B) \in \mathcal{L}^{p,\kappa}(\mathbb{G}^n) \times \mathcal{B}_a,$$

then

(2.9)
$$\left(V_\gamma(B)^{\kappa-1} \int_B |f|^p \, dV_\gamma \right)^{\frac{1}{p}} \lesssim \|f\|_{\mathcal{L}_{\mathcal{B}_a}^{p,\kappa}(\mathbb{G}^n)}.$$

We prove (2.9) by distinguishing the following two cases.

$Case \ |c_B| \leq 1 + a.$ Under this situation we have

$$\left(V_\gamma(B)^{\kappa-1} \int_B |f|^p \, dV_\gamma \right)^{\frac{1}{p}}$$

$$\leq \left(V_\gamma(B)^{\kappa-1} \int_B |f - f_{B,\gamma}|^p \, dV_\gamma \right)^{\frac{1}{p}} + V_\gamma(B)^{\frac{\kappa}{p}} |f_{B,\gamma}|$$

$$\leq \|f\|_{C_{\mathcal{B}_a}^{p,\kappa}(\mathbb{G}^n)} + V_\gamma(B)^{\frac{\kappa}{p}-1} \int_B |f - f_{B(0,1),\gamma}| \, dV_\gamma + V_\gamma(B)^{\frac{\kappa}{p}} |f_{B(0,1),\gamma}|$$

$$\lesssim \|f\|_{C_{\mathcal{B}_a}^{p,\kappa}(\mathbb{G}^n)} + \|f\|_{L^1(\mathbb{G}^n)}.$$

$Case \ |c_B| > 1 + a.$ Under this situation, we recall the notion of maximal admissible balls and the mother of a maximal admissible ball introduced in [40].

Given $a \in (0, \infty)$. A ball $B \in \mathcal{B}_a$ is said to be maximal if $r_B = a \, m(c_B)$. For each maximal ball $B \in \mathcal{B}_a$ containing no the origin, we denote by $M(B)$ the maximal ball in $\mathcal{B}_a$ centered at a point in the segment

$$[0, c_B] = \{ t c_B \ : \ t \in [0, 1] \}$$

such that the boundary of $M(B)$ contains c_B, and naturally we call $M(B)$ the mother of B. In other words, the relation between B and its mother $M(B)$ is as follows:

$$\begin{cases} r_{M(B)} = a \, m(c_{M(B)}); \\ |c_{M(B)}| + r_{M(B)} = |c_B|; \\ c_{M(B)} = \left(\frac{|c_{M(B)}|}{|c_B|} \right) c_B. \end{cases}$$

For notational convenience put

$$M^0(B) := B.$$

If $M(B)$ does not contain the origin, then we may consider the mother of $M(B)$, which is written as $M^2(B)$. Therefore, for any maximal ball $B \in \mathcal{B}_a$, we may find a chain of maximal balls,

$$B, \ M(B), \ M^2(B), \dots, M^k(B),$$

with the property that $M^j(B)$ is the mother of

$$M^{j-1}(B) \ \forall \ j \in \{1, \dots, k\}$$

and $M^k(B)$ contains the origin.

Now, we may assume that

$$\{M^k(B) : 0 \le k \le k_0\}$$

is a chain of maximal balls in $\mathcal{B}_a$, with the property that $M^k(B)$ is the mother of

$$M^{k-1}(B) \ \forall \ k \in \{1, \ldots, k_0\},$$

where k_0 is the smallest number such that

$$|c_{M^{k_0}(B)}| \le 1 + a.$$

For the notational convenience let

$$B_0 := B \ \& \ B_k := M^k(B) \ \forall \ k \in \{1, \ldots, k_0\}.$$

Then

$$\left(V_\gamma(B)^{\kappa-1} \int_B |f|^p \, dV_\gamma\right)^{\frac{1}{p}} \le I_1 + I_2 + I_3,$$

where

$$\begin{cases} I_1 := V_\gamma(B)^{\frac{\kappa}{p}} \left(\frac{1}{V_\gamma(B)} \int_B |f - f_{B,\gamma}|^p \, dV_\gamma\right)^{\frac{1}{p}}; \\ I_2 := V_\gamma(B)^{\frac{\kappa}{p}} \sum_{k=1}^{k_0} |f_{B_{k-1},\gamma} - f_{B_k,\gamma}|; \\ I_3 := V_\gamma(B)^{\frac{\kappa}{p}} |f_{B_{k_0},\gamma}|. \end{cases}$$

Obviously,

$$I_1 \le \|f\|_{C_{\mathcal{B}_a}^{p,\kappa}(\mathbb{G}^n)}.$$

For term I_3, we use

$$|c_{M^{k_0}(B)}| \le 1 + a$$

to get

$$V_\gamma(B_{k_0}) = V_\gamma\left(M^{k_0}(B)\right) \simeq 1,$$

which, combined with $V_\gamma(B) < 1$, yields

$$I_3 \le \frac{V_\gamma(B)^{\frac{\kappa}{p}}}{V_\gamma(B_{k_0})} \int_{B_{k_0}} |f| \, dV_\gamma \lesssim \|f\|_{L^1(\mathbb{G}^n)}.$$

To estimate I_2, for $1 \le k \le k_0$, since B_k is the mother of B_{k-1}, we use [36, Lemma 3.2] to get

$$\begin{cases} B_{k-1} \subseteq (a + 2) B_k; \\ V_\gamma(B_{k-1}) \simeq V_\gamma(B_k) \simeq V_\gamma((a+2)B_k), \end{cases}$$

which, together with the Hölder inequality, implies

$$
I_2 \leq V_\gamma(B)^{\frac{\kappa}{p}} \sum_{k=1}^{k_0} \left(|f_{B_{k-1},\gamma} - f_{(a+2)B_k,\gamma}| + |f_{(a+2)B_k,\gamma} - f_{B_k,\gamma}| \right)
$$

$$
\leq V_\gamma(B)^{\frac{\kappa}{p}} \sum_{k=1}^{k_0} \left(\frac{1}{V_\gamma(B_{k-1})} \int_{B_{k-1}} |f - f_{(a+2)B_k,\gamma}| \, dV_\gamma + \frac{1}{V_\gamma(B_k)} \int_{B_k} \cdots \right)
$$

$$
\leq \|f\|_{C_{\mathcal{B}a}^{p,\kappa}(\mathbb{G}^n)} \sum_{k=1}^{k_0} \left(\frac{V_\gamma(B)}{V_\gamma((a+2)B_k)} \right)^{\frac{\kappa}{p}} \left(\frac{V_\gamma((a+2)B_k)}{V_\gamma(B_{k-1})} + \frac{V_\gamma((a+2)B_k)}{V_\gamma(B_k)} \right)
$$

$$
\lesssim \|f\|_{C_{\mathcal{B}a}^{p,\kappa}(\mathbb{G}^n)} \sum_{k=1}^{k_0} \left(\frac{V_\gamma(B)}{V_\gamma(B_k)} \right)^{\frac{\kappa}{p}}.
$$

Note that (cf. [36, Lemma 3.3(iv)])

$$
V_\gamma(B) \lesssim e^{-ak} V_\gamma(B_k) \quad \forall \quad k \in \{1, \ldots, k_0\}.
$$

So we continuously deduce

$$
I_2 \lesssim \|f\|_{C_{\mathcal{B}a}^{p,\kappa}(\mathbb{G}^n)} \sum_{k=1}^{k_0} e^{-\frac{ak\kappa}{p}} \lesssim \|f\|_{C_{\mathcal{B}a}^{p,\kappa}(\mathbb{G}^n)}.
$$

Via combining the estimates for I_1, I_2 and I_3, we achieve the desired inequality (2.9).

(ii) This follows from the definition.

(iii) For $\kappa \in [-\frac{p}{n}, 0)$, it is easy to see that

$$
\mathrm{Lip}_{-\frac{\kappa}{p}}(\mathbb{G}^n) \subseteq C^{p,\kappa}(\mathbb{G}^n)
$$

by their definitions. Also, when $\kappa \in (-\infty, -\frac{p}{n})$, it is obvious that almost everywhere constant functions belong to $C^{p,\kappa}(\mathbb{G}^n)$.

Let

$$
f \in C^{p,\kappa}(\mathbb{G}^n) \quad \& \quad \kappa \in (-\infty, 0).
$$

For any $B \in \mathcal{B}_1$ and almost every $x, y \in B$, by the differential theorem of integrals, we obtain

$$
\begin{aligned}
|f(x) - f(y)| &= \lim_{j \to \infty} |f_{B(x, 2^{-j}r_B),\gamma} - f_{B(y, 2^{-j}r_B),\gamma}| \\
(2.10) \qquad &\leq \lim_{j \to \infty} |f_{B(x, 2^{-j}r_B),\gamma} - f_{B(x, r_B),\gamma}| + f_{B(x, r_B),\gamma} - f_{B(y, r_B),\gamma}| \\
&\quad + \lim_{j \to \infty} |f_{B(y, r_B),\gamma} - f_{B(y, 2^{-j}r_B),\gamma}|.
\end{aligned}
$$

From Lemma 2.1.2 and

$$V_\gamma(B(x, 2^{-j}r_B)) \simeq 2^{-jn}V_\gamma(B),$$

we deduce

(2.11) $\quad \left| f_{B(x,2^{-j}r_B),\gamma} - f_{B(x,r_B),\gamma} \right| \lesssim \dfrac{2^{-\frac{jn\kappa}{p}}\|f\|_{C^{p,\kappa}(\mathbb{G}^n)}}{V_\gamma(B(x,2^{-j}r_B))^{\frac{\kappa}{p}}} \lesssim \dfrac{\|f\|_{C^{p,\kappa}(\mathbb{G}^n)}}{V_\gamma(B)^{\frac{\kappa}{p}}}.$

Likewise, for almost every $y \in B$, we have

(2.12) $\qquad \left| f_{B(y,2^{-j}r_B),\gamma} - f_{B(y,r_B),\gamma} \right| \lesssim \dfrac{\|f\|_{C^{p,\kappa}(\mathbb{G}^n)}}{V_\gamma(B)^{\frac{\kappa}{p}}}.$

For all $x, y \in B$, we observe

$$B(x, r_B) \subseteq 2B \;\;\&\;\; B(y, r_B) \subseteq 2B$$

whence, by (2.2) and (2.3),

$$V_\gamma(B) \simeq V_\gamma\big(B(x, r_B)\big) \simeq V_\gamma\big(B(y, r_B)\big).$$

From these and Lemma 2.1.2, it follows that

(2.13)
$$\begin{aligned} \left| f_{B(x,r_B),\gamma} - f_{B(y,r_B),\gamma} \right| & \\ \le \left| f_{B(x,r_B),\gamma} - f_{2B,\gamma} \right| &+ \left| f_{2B,\gamma} - f_{B(y,r_B),\gamma} \right| \\ \lesssim \dfrac{\|f\|_{C^{p,\kappa}(\mathbb{G}^n)}}{V_\gamma(B)^{\frac{\kappa}{p}}}. & \end{aligned}$$

Inserting (2.11), (2.12), and (2.13) into (2.10), and using (2.3), we see that for almost every

$$(x, y) \in B \times B \in \mathcal{B}_a \times \mathcal{B}_a$$

there holds

(2.14) $\qquad |f(x) - f(y)| \lesssim \dfrac{\|f\|_{C^{p,\kappa}(\mathbb{G}^n)}}{V_\gamma(B)^{\frac{\kappa}{p}}},$

whence finding

$$f \in \text{Lip}_{-\frac{\kappa}{p}}(\mathbb{G}^n) \; \forall \; \kappa \in \left[-\frac{p}{n}, 0 \right).$$

(iv) When $\kappa \in (-\infty, -\frac{p}{n})$, from (2.14) one deduces that the derivative of f is 0 for almost every $x \in B$, and hence f is a constant function almost everywhere. $\square$

2.2 Another Look at $C^{p,\kappa}(\mathbb{G}^n)$ for $-\frac{p}{n} \leq \kappa < 0$

As well known (cf. [51]), the Ornstein-Uhlenbeck-Poisson semigroup $\{\mathcal{P}_t\}_{t>0}$ is determined by

$$\mathcal{P}_t f(x) = \int_{\mathbb{R}^n} \mathcal{P}_t(x, y) f(y) \, dV(y)$$

where

$$\mathcal{P}_t(x, y) = \frac{1}{2\pi^{\frac{n+1}{2}}} \int_0^\infty \left(\frac{t}{s^{\frac{3}{2}}}\right) e^{-\frac{t^2}{4s}} \left(\frac{e^{-\frac{|y-e^{-s}x|^2}{1-e^{-2s}}}}{(1-e^{-2s})^{\frac{n}{2}}}\right) ds$$

is valid for any

$$(t, x, y) \in (0, \infty) \times \mathbb{R}^n \times \mathbb{R}^n.$$

In accordance with [23, 24, 46], we say that for $0 < \alpha < 1$ a function f on $\mathbb{R}^n$ is said to be in $Lip_{b,\alpha}(\mathbb{G}^n)$ provided

$$\|f\|_{Lip_{b,\alpha}(\mathbb{G}^n)} := \|f\|_{L^\infty(\mathbb{R}^n)} + \sup_{t \in (0,\infty)} t^{1-\alpha} \|\partial_t \mathcal{P}_t f\|_{L^\infty(\mathbb{R}^n)} < \infty.$$

Lemma 2.2.1. Let $\alpha \in (0, 1)$. Then $f \in Lip_{b,\alpha}(\mathbb{G}^n)$ if and only if

$$(2.15) \qquad |f(x) - f(y)| \lesssim \min\left\{|x - y|^\alpha, \left(\frac{|x - y_x|}{1 + |x|}\right)^{\frac{\alpha}{2}} + |y'_x|^\alpha\right\} \ \forall \ x, y \in \mathbb{R}^n$$

after correction of f on a null set. Here and hereafter, for any $x, y \in \mathbb{R}^n$ with $x \neq 0$, we decompose y as $y = y_x + y'_x$, where y_x is parallel to x and y'_x orthogonal to x; if $x = 0$, then $y_x = y$ and $y'_x = 0$, and this also holds for all x in case $n = 1$. Moreover,

$$\|f\|_{Lip_{b,\alpha}(\mathbb{G}^n)} \simeq \|f\|_{L^\infty(\mathbb{R}^n)} + \inf \{C > 0 : \text{ constants C in } (2.15)\}$$

$$\simeq |f(0)| + \inf \{C > 0 : \text{ constants C in } (2.15)\}$$

and the implicit equivalent constants are independent of f.

Proof. Note that this lemma means that the combined Lipschitz condition applies in the radial direction, but in the orthogonal direction the exponent is always α. So this lemma follows from [35, Theorem 1.1] and the fact that (2.15) amounts to

$$|f(x) - f(y)| \lesssim \min\left\{|x - y|^\alpha, \left(\frac{|x - y|}{1 + |x| + |y|}\right)^{\frac{\alpha}{2}} + \left(\frac{\sin \theta}{(|x| + |y|)^{-1}}\right)^\alpha\right\} \ \forall \ x, y \in \mathbb{R}^n$$

after correction of f on a null set; where θ is the angle between x and y and vanishes whenever x or y is the origin, along with a combination of the ordinary Lipschitz continuity conditions (cf. [52, Section V.4.2]), some sharp estimates for the Ornstein-Uhlenbeck-Poisson semigroup $(\mathcal{P}_t)_{t \geq 0}$ and some of its derivatives. $\quad\square$

From the above characterization of the Gaussian Lipschitz space, we know that any

$$f \in Lip_{b,\alpha}(\mathbb{G}^n) \quad \text{with} \quad \alpha \in (0,1)$$

behaves like the classical Lipschitz function when x is close to y (namely, $|x - y|$ is smaller than a fixed constant multiple of $(1 + |x|)^{-1}$); and the combined Lipschitz condition plays a role when $|x - y|$ is large but not "quite" large - in other words - $|x - y|$ is smaller than a fixed constant; for the remaining case the fact

$$f \in L^\infty(\mathbb{R}^n)$$

will be important.

Based on these, for any

$$(x, r) \in \mathbb{R}^n \times \left(\frac{1}{1 + |x|}, 1 \right]$$

we define the Gaussian cylinder

$$R(x, r) := \left\{ y \in \mathbb{R}^n : |y_x - x| < \ell = 2^{-1} r^2 (1 + |x|) \ \& \ |y_x'| < r \right\}.$$

Notice that $R(x, r)$ is a cylinder centered at x, with bottom radius r and length 2ℓ, but it goes along the direction x.

Definition 2.2.2. For $\alpha \in (0, 1)$ and $p \in [1, \infty)$ let $Lip_{\alpha,p}(\mathbb{G}^n)$ be the collection of all locally integrable functions f on $\mathbb{R}^n$ such that

$$\|f\|_{Lip_{\alpha,p}(\mathbb{G}^n)} := \sup_{(x,r) \in \mathbb{R}^n \times (\frac{1}{1+|x|}, 1]} r^{-\alpha} \left(\frac{1}{V(R(x,r))} \int_{R(x,r)} |f - f_{R(x,r)}|^p \, dV \right)^{\frac{1}{p}}$$

$$+ \sup_{(x,r) \in \mathbb{R}^n \times (0, 1]} r^{-\alpha} \left(\frac{1}{V(B(x,r))} \int_{B(x,r)} |f - f_{B(x,r)}|^p \, dV \right)^{\frac{1}{p}}$$

$$+ \sup_{(x,r) \in \mathbb{R}^n \times (1, \infty)} r^{-\alpha} \left(\frac{1}{V(B(x,r))} \int_{B(x,r)} |f|^p \, dV \right)^{\frac{1}{p}}$$

is finite. Here and hereafter in this section, for any set $E \subseteq \mathbb{R}^n$ we make a convention

$$f_E = \frac{1}{V(E)} \int_E f \, dV.$$

Another perspective of the formula

$$C^{p,\kappa}(\mathbb{G}^n) = Lip_{-\frac{\kappa}{p}}(\mathbb{G}^n) \quad \forall \quad k \in \left[-\frac{p}{n}, 0 \right)$$

is the following assertion.

Theorem 2.2.3. Let $\alpha \in (0, 1)$ and $p \in [1, \infty)$. Then

$$Lip_{\alpha,p}(\mathbb{G}^n) = Lip_{b,\alpha}(\mathbb{G}^n)$$

holds with equivalent norms.

Proof. The required identification follows from a consideration of two parts.

$$\text{Part 1}: \quad f \in Lip_{\alpha,p}(\mathbb{G}^n) \Leftarrow f \in Lip_{b,\alpha}(\mathbb{G}^n).$$

Let

$$\|f\|_{Lip_{b,\alpha}(\mathbb{G}^n)} = 1.$$

By Lemma 2.2.1, after correction of f on a null set, one has that if $x, z \in \mathbb{R}^n$ then

(2.16)
$$|f(x) - f(z)| \lesssim |x - z|^\alpha$$

and

(2.17)
$$|f(x) - f(z)| \lesssim \left(\frac{|x - z_x|}{1 + |x|}\right)^{\frac{\alpha}{2}} + |z_x'|^\alpha.$$

For any

$$(x, r) \in \mathbb{R}^n \times \left(\frac{1}{1 + |x|}, 1\right] \quad \& \quad z \in R(x, r),$$

we have

$$|z_x'| < r \quad \& \quad |z_x - x| < 2^{-1} r^2 (1 + |x|),$$

so that (2.17) implies

$$|f(z) - f(x)| \lesssim r^\alpha.$$

Consequently,

$$\left(\frac{1}{V(R(x,r))} \int_{R(x,r)} |f - f_{R(x,r)}|^p \, dV\right)^{\frac{1}{p}}$$

$$\leq 2 \left(\frac{1}{V(R(x,r))} \int_{R(x,r)} |f - f(x)|^p \, dV\right)^{\frac{1}{p}}$$

$$\lesssim r^\alpha.$$

Similarly, for any

$$(x, r) \in \mathbb{R}^n \times (0, 1] \quad \& \quad z \in B(x, r),$$

the estimation (2.16) also implies

$$|f(z) - f(x)| \lesssim r^\alpha,$$

whence leading to

$$\left(\frac{1}{V(B(x,r))} \int_{B(x,r)} |f - f_{B(x,r)}|^p \, dV\right)^{\frac{1}{p}}$$

$$\leq 2 \left(\frac{1}{V(B(x,r))} \int_{B(x,r)} |f - f(x)|^p \, dV \right)^{\frac{1}{p}}$$
$$\lesssim r^\alpha.$$

Also, for any

$$(x,r) \in \mathbb{R}^n \times (1,\infty)$$

we utilize

$$f \in L^\infty(\mathbb{R}^n)$$

to estimate

$$r^{-\alpha} \left(\frac{1}{V(B(x,r))} \int_{B(x,r)} |f|^p \, dV \right)^{\frac{1}{p}} \lesssim \|f\|_{L^\infty(\mathbb{R}^n)}.$$

Combining the last three formulae gives

$$\|f\|_{\mathrm{Lip}_{a,p}(\mathbb{G}^n)} \lesssim 1 \simeq \|f\|_{\mathrm{Lip}_{b,\alpha}(\mathbb{G}^n)}.$$

$$\textit{Part 2}: \quad f \in \mathrm{Lip}_{a,p}(\mathbb{G}^n) \Rightarrow f \in \mathrm{Lip}_{b,\alpha}(\mathbb{G}^n).$$

Let

$$\|f\|_{\mathrm{Lip}_{a,p}(\mathbb{G}^n)} = 1.$$

We are about to show

$$\|f\|_{\mathrm{Lip}_{b,\alpha}(\mathbb{G}^n)} \lesssim 1.$$

From the definition of $\mathrm{Lip}_{a,p}(\mathbb{G}^n)$ it follows that

$$\sup_{(x,r) \in \mathbb{R}^n \times (0,\infty)} r^{-\alpha} \left(\frac{1}{V(B(x,r))} \int_{B(x,r)} |f - f_{B(x,r)}|^p \, dV \right)^{\frac{1}{p}} \lesssim 1,$$

which, together with a standard classical argument, implies that for almost all $x, y \in \mathbb{R}^n$ one has

(2.18)
$$|f(x) - f(y)| \lesssim |x - y|^\alpha.$$

It remains to show that for almost all $x, y \in \mathbb{R}^n$ one has

(2.19)
$$|f(x) - f(y)| \lesssim \left(\frac{|x - y_x|}{1 + |x|} \right)^{\frac{\alpha}{2}} + |y'_x|^\alpha.$$

Via writing

$$|f(x) - f(y)| \leq |f(x) - f(y_x)| + |f(y_x) - f(y)|$$

and noticing

$$|f(y_x) - f(y)| \lesssim |y_x - y|^\alpha \simeq |y'_x|^\alpha$$

by means of (2.18), we observe that the proof of (2.19) can be reduced to proving

$$|f(x) - f(y_x)| \lesssim \left(\frac{|x - y_x|}{1 + |x|} \right)^{\frac{\alpha}{2}}.$$

In other words, we only need to prove that if

$$(x, y) \in \mathbb{R}^n \times \mathbb{R}^n \text{ with } x = \tau y \text{ for some } \tau \in (-\infty, \infty)$$

then

(2.20)
$$|f(x) - f(y)| \lesssim \left(\frac{|x - y|}{1 + |x|} \right)^{\frac{\alpha}{2}}.$$

The demonstration of (2.20) can be split into the following three cases.

$$Case \ i = \begin{cases} |x - y| \leq \frac{10^{10}}{1+|x|} & \text{as } i = 1; \\ \frac{10^{10}}{1+|x|} < |x - y| < \frac{1+|x|}{10^{10}} & \text{as } i = 2; \\ |x - y| \geq \frac{1+|x|}{10^{10}} & \text{as } i = 3. \end{cases}$$

For *Case 1* we use

$$|x - y| \leq \frac{10^{10}}{1 + |x|}$$

and (2.18) to get the required inequality

$$|f(x) - f(y)| \lesssim |x - y|^\alpha \lesssim \left(\frac{|x - y|}{1 + |x|} \right)^{\frac{\alpha}{2}}.$$

For *Case 2* we have

$$\frac{10^{10}}{1 + |x|} < |x - y| < \frac{1 + |x|}{10^{10}},$$

whence

$$|x| \geq 10^9 \ \& \ 2^{-1}|y| < |x| < 2|y|.$$

For each $j + 1 \in \mathbb{N}$ let

$$\begin{cases} r_j = \sqrt{\frac{2^{2-j}|x-y|}{1+|x|}} & \& \quad R_j := R(x, r_j); \\ r'_j = \sqrt{\frac{2^{-j}|x-y|}{1+|y|}} & \& \quad R'_j := R(y, r'_j), \end{cases}$$

and observe

$$\begin{cases} V(R_j) \simeq r_j^{n+1}(1 + |x|) \simeq V(R_{j+1}); \\ V(R'_j) \simeq (r'_j)^{n+1}(1 + |y|) \simeq V(R'_{j+1}). \end{cases}$$

From the last two estimates and the equivalence

$$1 + |x| \simeq 1 + |y|$$

it follows that

$$V(R'_1) \simeq V(R_0).$$

Now, we show $R'_1 \subseteq R_0$. To this end, let $z \in R'_1$. Then

$$|z'_y| < r \ \ \& \ \ |z_y - y| < 2^{-1}(r'_1)^2(1 + |y|) = 2^{-2}|x - y|.$$

Using the facts

$$2^{-1}|y| < |x| < 2|y| \ \ \text{and} \ \ x = \tau y \ \ \text{for some} \ \ \tau \in (-\infty, \infty),$$

we obtain

$$|z'_x| = |z'_y| < r$$

and

$$|z_x - x| = \frac{|z_y - y||x|}{|y|} < \frac{|x - y||x|}{4|y|} < 2|x - y| = 2^{-1}r_0^2(1 + |x|),$$

which implies $z \in R_0$ and proves $R'_1 \subseteq R_0$. By the Lebesgue differential theorem of integrals we obtain

$$(2.21) \qquad |f(x) - f(y)| \le \sum_{j=1}^{\infty} \left| f_{R_j} - f_{R_{j-1}} \right| + \left| f_{R_0} - f_{R'_1} \right| + \sum_{j=1}^{\infty} \left| f_{R'_j} - f_{R'_{j+1}} \right|.$$

Upon using

$$V(R_j) \simeq V(R_{j-1}) \ \forall \ j \in \mathbb{N},$$

we gain

$$\left| f_{R_j} - f_{R_{j-1}} \right| \le \left(\frac{1}{V(R_j)} \int_{R_j} \left| f - f_{R_{j-1}} \right|^p dV \right)^{\frac{1}{p}}$$

$$\le \left(\frac{1}{V(R_{j-1})} \int_{R_{j-1}} \left| f - f_{R_{j-1}} \right|^p dV \right)^{\frac{1}{p}}$$

$$\le (r_{j-1})^{\alpha}$$

$$\simeq \left(\frac{2^{-j}|x - y|}{1 + |x|} \right)^{\frac{\alpha}{2}}.$$

Similarly,

$$\left| f_{R'_j} - f_{R'_{j+1}} \right| \lesssim (r'_j)^{\alpha} \simeq \left(\frac{2^{-j}|x - y|}{1 + |y|} \right)^{\frac{\alpha}{2}} \simeq \left(\frac{2^{-j}|x - y|}{1 + |x|} \right)^{\frac{\alpha}{2}}.$$

Applying

$$R'_1 \subseteq R_0 \ \ \& \ \ V(R'_1) \simeq V(R_0),$$

we derive

$$\left| f_{R_0} - f_{R'_1} \right| \le \left(\frac{1}{V(R'_1)} \int_{R'_1} \left| f - f_{R_0} \right|^p dV \right)^{\frac{1}{p}}$$

$$\lesssim \left(\frac{1}{V(R_0)} \int_{R_0} |f - f_{R_0}|^p \, dV \right)^{\frac{1}{p}}$$

$$\lesssim (r_0)^\alpha$$

$$\simeq \left(\frac{|x - y|}{1 + |x|} \right)^{\frac{\alpha}{2}}.$$

Inserting the last three formulae in (2.21) leads to (2.20).

For *Case 3* we utilize

$$|x - y| \geq (1 + |x|) 10^{-10}$$

to get that if

$$z \in B(x, 10^{-10})$$

then

$$\max\{|z_x - x|, |z'_x|\} \leq |z - x| < 10^{-10} < (1 + |x|) 10^{-10}$$

and hence

$$|f(z) - f(x)| \leq |f(z_x) - f(x)| + |f(z_x) - f(z)|$$

$$\lesssim |z'_x|^\alpha + \left(\frac{|z_x - x|}{1 + |x|} \right)^{\frac{\alpha}{2}}$$

$$\lesssim 1,$$

where we have used the already-established (2.18) to estimate

$$|f(z_x) - f(z)|,$$

and the already-proved results in *Case 1* and *Case 2* to control

$$|f(z_x) - f(x)|.$$

Thus

$$\left| f(x) - f_{B(x, 10^{-10})} \right| \leq \left(\frac{1}{V(B(x, 10^{-10}))} \int_{B(x, 10^{-10})} |f - f(x)|^p \, dV \right)^{\frac{1}{p}} \lesssim 1.$$

Meanwhile, the definition of $\mathrm{Lip}_{\alpha, p}$ implies

$$\left| f_{B(x, 10^{-10})} \right| \lesssim \left(\frac{1}{V(B(x, 1))} \int_{B(x, 1)} |f|^p \, dV \right)^{\frac{1}{p}} \lesssim 1.$$

Combining the last two formulae implies

$$|f(x)| \leq \left| f(x) - f_{B(x, 10^{-10})} \right| + \left| f_{B(x, 10^{-10})} \right| \lesssim 1.$$

Analogously, we have

$$\begin{cases} |f(y)| \lesssim 1; \\ |f(x) - f(y)| \lesssim 1 \lesssim \left(\frac{|x-y_x|}{1+|x|}\right)^{\frac{\alpha}{2}}, \end{cases}$$

thereby reaching (2.20) and completing the proof of *Part 2*.

$\square$

Corollary 2.2.4. Let $\alpha \in (0, 1)$ and $p \in [1, \infty)$. Then one has the Gaussian Campanato estimation

$$\|f\|_{Lip_{b,\alpha}(\mathbb{G}^n)} \simeq \|f\|_{L^\infty(\mathbb{R}^n)}$$

$$+ \sup_{(x,r)\in\mathbb{R}^n\times\left(\frac{1}{1+|x|}, 1\right]} r^{-\alpha} \left(\frac{1}{V(R(x,r))} \int_{R(x,r)} |f - f_{R(x,r)}|^p \, dV\right)^{\frac{1}{p}}$$

$$+ \sup_{(x,r)\in\mathbb{R}^n\times(0, 1]} r^{-\alpha} \left(\frac{1}{V(B(x,r))} \int_{B(x,r)} |f - f_{B(x,r)}|^p \, dV\right)^{\frac{1}{p}},$$

thereby finding the Gaussian Morrey inequality

$$\|f\|_{Lip_{b,\alpha}(\mathbb{G}^n)} \lesssim \|f\|_{L^\infty(\mathbb{R}^n)}$$

$$+ \sup_{(x,r)\in\mathbb{R}^n\times\left(\frac{1}{1+|x|}, 1\right]} r^{\alpha} \left(\frac{1}{V(R(x,r))} \int_{R(x,r)} |\nabla f|^p \, dV\right)^{\frac{1}{p}}$$

$$+ \sup_{(x,r)\in\mathbb{R}^n\times(0, 1]} r^{-\alpha} \left(\frac{1}{V(B(x,r))} \int_{B(x,r)} |\nabla f|^p \, dV\right)^{\frac{1}{p}}.$$

Proof. This follows from the argument for Theorem 2.2.3 and the Poincaré inequality:

$$\left(|f - f_E|^p\right)_E \lesssim \left(|\nabla f|^p\right)_E \quad \text{as} \quad E = R(x,r) \quad \text{or} \quad E = B(x,r).$$

$\square$

Chapter 3
Gaussian p-Capacity

In this chapter we introduce the notion of the Gaussian p-capacity which is important at least in describing the appropriate tracing of the Gaussian Sobolev p-space.

3.1 Gaussian p-Capacity for $1 \le p < \infty$

Definition 3.1.1. For $p \in [1, \infty)$ and $E \subseteq \mathbb{R}^n$ let

$$\mathcal{A}_p(E) := \left\{ f \in W^{1,p}(\mathbb{G}^n) : \ E \subseteq \{x \in \mathbb{R}^n : f(x) \ge 1\}^\circ \right\}.$$

Define the Gaussian p-capacity of E as:

$$(3.1) \qquad \mathrm{Cap}_p(E; \mathbb{G}^n) := \inf \left\{ \|f\|^p_{W^{1,p}(\mathbb{G}^n)} : \ f \in \mathcal{A}_p(E) \right\}.$$

Recall that $C_c^1(\mathbb{R}^n; \mathbb{R}^n)$ is the collection of all vector-valued functions

$$\Phi = (\phi_1, \ldots, \phi_n) \quad \text{with} \quad \phi_i \in C_c^1(\mathbb{R}^n).$$

Similarly, $C_c^0(\mathbb{R}^n; \mathbb{R}^n)$ is the collection of all vector-valued functions

$$\Phi = (\phi_1, \ldots, \phi_n) \quad \text{with} \quad \phi_i \in C_c^0(\mathbb{R}^n).$$

Lemma 3.1.2. Given $p \in [1, \infty)$. For a sequence of functions

$$\{f_k\}_{k \in \mathbb{N}} \subseteq W^{1,p}(\mathbb{G}^n)$$

let

$$g = \sup_{k \in \mathbb{N}} f_k \quad \text{and} \quad h = \sup_{k \in \mathbb{N}} |\nabla f_k|.$$

If both g and h are in $L^1(\mathbb{G}^n)$, then

$$|\nabla g(x)| \le h(x)$$

holds for almost all $x \in \mathbb{R}^n$.

Proof. Thanks to

$$g \in L^1(\mathbb{G}^n) \subseteq L^1_{\mathrm{loc}}(\mathbb{R}^n),$$

© Springer Nature Switzerland AG 2018
L. Liu et al., *Gaussian Capacity Analysis*, Lecture Notes in Mathematics 2225,
https://doi.org/10.1007/978-3-319-95040-2_3

it makes sense to consider ∇g. According to (1.1), we have

$$\int_{\mathbb{R}^n} \nabla g \cdot \Phi \, dV = -\int_{\mathbb{R}^n} g \, \mathrm{div} \Phi \, dV \quad \forall \quad \Phi \in C_c^1(\mathbb{R}^n; \mathbb{R}^n).$$

For any $l \in \mathbb{N}$, define

$$g_l = \sup_{1 \le k \le l} f_k.$$

Due to

$$f_k \in W^{1,p}(\mathbb{G}^n),$$

one has

$$g_l \in L^p(\mathbb{G}^n) \subseteq L^1_{\mathrm{loc}}(\mathbb{R}^n).$$

An application of [18, p. 148, Lemma 2(iii)] yields

(3.2)
$$|\nabla g_l| \le \sup_{1 \le k \le l} |\nabla f_k| \quad \text{a.e. on} \quad \mathbb{R}^n.$$

Of course, this follows by induction and from verifying the case $l = 2$:

$$\nabla \max\{f_1, f_2\}(x) = 2^{-1}\Big(\nabla f_1(x) + \nabla f_2(x) + \nabla|f_1(x) - f_2(x)|\Big)$$

$$= \begin{cases} \nabla f_1(x) & \text{for almost all } x \in \mathbb{R}^n \text{ with } f_1(x) \ge f_2(x); \\ \nabla f_2(x) & \text{for almost all } x \in \mathbb{R}^n \text{ with } f_1(x) \le f_2(x). \end{cases}$$

According to (1.1) and the Lebesgue dominated convergence theorem, for any

$$\Phi \in C_c^1(\mathbb{R}^n; \mathbb{R}^n),$$

we have

$$\left| \int_{\mathbb{R}^n} g \, \mathrm{div} \Phi \, dV \right| = \left| \lim_{l \to \infty} \int_{\mathbb{R}^n} g_l \, \mathrm{div} \Phi \, dV \right|$$

$$= \left| \lim_{l \to \infty} \int_{\mathbb{R}^n} \nabla g_l \cdot \Phi \, dV \right|$$

$$\le \int_{\mathbb{R}^n} |\Phi| h \, dV.$$

Based on this and Rudin [49, p. 58, Theorem 3.3], the linear functional L defined by

$$L(\Phi) := \int_{\mathbb{R}^n} g \, \mathrm{div} \Phi \, dV \quad \forall \quad \Phi \in C_c^1(\mathbb{R}^n; \mathbb{R}^n)$$

extends to a linear functional $\bar{L}$ on $C_c^0(\mathbb{R}^n; \mathbb{R}^n)$ such that

$$|\bar{L}(\Phi)| \le \int_{\mathbb{R}^n} |\Phi| h \, dV \quad \forall \quad \Phi \in C_c^0(\mathbb{R}^n; \mathbb{R}^n).$$

In particular, for each compact set $K \subseteq \mathbb{R}^n$,

$$\sup \left\{ \bar{L}(\Phi) : \ \Phi \in C_c^0(\mathbb{R}^n; \mathbb{R}^n), \ |\Phi| \leq 1, \ \text{supp} \, \Phi \subseteq K \right\}$$
$$\leq \int_K h \, dV$$
$$\leq C(K, n) \|h\|_{L^1(\mathbb{G}^n)} < \infty.$$

By this and [18, p. 49, Theorem 1], there exist a Radon measure μ on $\mathbb{R}^n$ and a μ-measurable function $\sigma : \mathbb{R}^n \to \mathbb{R}^n$ such that

$$\bar{L}(\Phi) = \int_{\mathbb{R}^n} \Phi \cdot \sigma \, d\mu \quad \forall \quad \Phi \in C_c^0(\mathbb{R}^n; \mathbb{R}^n),$$

where $|\sigma(x)| = 1$ for almost all $x \in \mathbb{R}^n$. Moreover, the construction of μ (see [18, p. 49, Definition]) gives us that for any Lebesgue measurable set $A \subseteq \mathbb{R}^n$,

$$\mu(A) = \inf\{\mu(O) : \text{ open } O \supseteq A\}$$
$$= \inf_{\text{open } O \supseteq A} \sup \left\{ \bar{L}(\Phi) : \ \Phi \in C_c^0(\mathbb{R}^n; \mathbb{R}^n), \ |\Phi| \leq 1, \ \text{supp}(\Phi) \subseteq O \right\}$$
$$\leq \inf_{\text{open } O \supseteq A} \int_O h \, dV$$
$$= \int_A h \, dV.$$

Therefore, μ is absolutely continuous with respect to the Lebesgue measure, so that

$$d\mu = u \, dV$$

for some function u satisfying that

$$|u(x)| \leq h(x)$$

for almost all $x \in \mathbb{R}^n$. Hence, for all

$$\Phi \in C_c^1(\mathbb{R}^n; \mathbb{R}^n),$$

we have

$$\int_{\mathbb{R}^n} g \, \text{div} \Phi \, dV = L(\Phi) = \bar{L}(\Phi) = \int_{\mathbb{R}^n} \Phi \cdot \sigma u \, dV,$$

which implies

$$\nabla g = \sigma u$$

and

$$|\nabla g(x)| \leq h(x) \text{ for almost all } x \in \mathbb{R}^n.$$

This completes the argument. $\qquad \square$

Remark 3.1.3. Three comments are in order.

(i) Let $p \in [1, \infty)$. Because of Remark 1.1.2 and Corollary 1.1.4, we have

$$\mathrm{Cap}_p(E; \mathbb{G}^n) \simeq \inf \left\{ \left(\|\|\nabla f\|\|_{L^p(\mathbb{G}^n)} + \|f\|_{L^1(\mathbb{G}^n)} \right)^p : f \in \mathcal{A}_p(E) \right\}$$

$$\simeq \inf \left\{ \left(\|\|\nabla f\|\|_{L^p(\mathbb{G}^n)} + \left| \int_{\mathbb{R}^n} f \, dV_\gamma \right| \right)^p : f \in \mathcal{A}_p(E) \right\}.$$

(ii) Given any function

$$f \in W^{1,p}(\mathbb{G}^n) \text{ with } p \in [1, \infty),$$

since

$$|f| = \max\{f, -f\},$$

it follows from Lemma 3.1.2 that

$$|\nabla(|f|)(x)| \le |\nabla f(x)|$$

for almost all $x \in \mathbb{R}^n$, which consequently implies

$$\|\|f\|\|_{W^{1,p}(\mathbb{G}^n)} \le \|f\|_{W^{1,p}(\mathbb{G}^n)} \ \forall f \in W^{1,p}(\mathbb{G}^n).$$

Observe that

$$f \in \mathcal{A}_p(E) \Rightarrow |f| \in \mathcal{A}_p(E).$$

So

$$\mathrm{Cap}_p(E; \mathbb{G}^n) = \inf \left\{ \|f\|_{W^{1,p}(\mathbb{G}^n)}^p : f \ge 0, \ f \in \mathcal{A}_p(E) \right\}.$$

(iii) When

$$f \in \mathcal{A}_p(E) \text{ with } p \in [1, \infty),$$

one easily deduces from [18, p. 130, Theorem 4(iii)] that

$$|\nabla \min\{f, 1\}| \le |\nabla f| \text{ a. e. on } \mathbb{R}^n,$$

so that

$$\| \min\{f, 1\} \|_{W^{1,p}(\mathbb{G}^n)} \le \|f\|_{W^{1,p}(\mathbb{G}^n)}$$

and then

$$\min\{f, 1\} \in \mathcal{A}_p(E).$$

Thus, we can equivalently write

$$\mathrm{Cap}_p(E; \mathbb{G}^n) = \inf \left\{ \|f\|_{W^{1,p}(\mathbb{G}^n)}^p : 0 \le f \le 1, \ f \in \mathcal{A}_p(E) \right\}.$$

Proposition 3.1.4. Let $p \in [1, \infty)$. The set-function $\mathrm{Cap}_p(\cdot \, ; \ \mathbb{G}^n)$ enjoys the following properties.

(i) $\mathrm{Cap}_p(\emptyset \, ; \ \mathbb{G}^n) = 0$ and $\mathrm{Cap}_p(\mathbb{R}^n \, ; \ \mathbb{G}^n) \le 1$.

(ii) If $E_1 \subseteq E_2 \subseteq \mathbb{R}^n$, then $\mathrm{Cap}_p(E_1;\ \mathbb{G}^n) \le \mathrm{Cap}_p(E_2;\ \mathbb{G}^n)$.

(iii) For any sequence $\{E_j\}_{j=1}^\infty$ of subsets of $\mathbb{R}^n$,

$$(3.3) \qquad \mathrm{Cap}_p\left(\cup_{j=1}^\infty E_j;\ \mathbb{G}^n\right) \le \sum_{j=1}^\infty \mathrm{Cap}_p(E_j;\ \mathbb{G}^n).$$

(iv) For any $1 \le p < q < \infty$ and any set $E \subseteq \mathbb{R}^n$,

$$2^{-\frac{1}{p}}\left(\mathrm{Cap}_p(E;\ \mathbb{G}^n)\right)^{\frac{1}{p}} \le 2^{-\frac{1}{q}}\left(\mathrm{Cap}_q(E;\ \mathbb{G}^n)\right)^{\frac{1}{q}}.$$

(v) For any sequence $\{K_j\}_{j=1}^\infty$ of compact subsets of $\mathbb{R}^n$ such that

$$\begin{cases} K_1 \supseteq K_2 \supseteq \cdots ; \\ \lim_{j\to\infty} \mathrm{Cap}_p\left(K_j;\ \mathbb{G}^n\right) = \mathrm{Cap}_p\left(\cap_{j=1}^\infty K_j;\ \mathbb{G}^n\right). \end{cases}$$

Proof. (i) It is easy to see that

$$\mathrm{Cap}_p(\emptyset;\ \mathbb{G}^n) = 0$$

holds by using Definition 3.1.1. To see the inequality

$$\mathrm{Cap}_p(\mathbb{R}^n;\ \mathbb{G}^n) \le 1,$$

we can take the test function $f \equiv 1$ in Definition 3.1.1.
  (ii) For any sets

$$E_1 \subseteq E_2 \subseteq \mathbb{R}^n,$$

we have

$$\mathcal{A}_p(E_2) \subseteq \mathcal{A}_p(E_1),$$

so that

$$\mathrm{Cap}_p(E_1;\ \mathbb{G}^n) = \inf_{f \in \mathcal{A}_p(E_1)} \|f\|_{W^{1,p}(\mathbb{G}^n)}^p \le \inf_{f \in \mathcal{A}_p(E_2)} \|f\|_{W^{1,p}(\mathbb{G}^n)}^p = \mathrm{Cap}_p(E_2;\ \mathbb{G}^n).$$

  (iii) Without loss of generality, we may assume

$$\sum_{j=1}^\infty \mathrm{Cap}_p(E_j;\ \mathbb{G}^n) < \infty$$

- otherwise - (3.3) holds trivially.
  For any $\epsilon > 0$ and $j \in \mathbb{N}$, by (3.1), we find a function

$$f_{j,\epsilon} \in \mathcal{A}_p(E_j)$$

such that

$$\mathrm{Cap}_p(E_j;\ \mathbb{G}^n) \le \|f_{j,\epsilon}\|_{W^{1,p}(\mathbb{G}^n)}^p \le \mathrm{Cap}_p(E_j;\ \mathbb{G}^n) + 2^{-j}\epsilon.$$

Let
$$f_\epsilon = \sup_{j \in \mathbb{N}} f_{j,\epsilon}.$$

Clearly,
$$\cup_{j=1}^{\infty} E_j \subseteq \{x \in \mathbb{R}^n : f_\epsilon(x) \ge 1\}^\circ.$$

Upon observing

$$
\begin{aligned}
(3.4) \qquad & \left\| \sup_{j \in \mathbb{N}} |\nabla f_{j,\epsilon}| \right\|_{L^p(\mathbb{G}^n)}^p + \left\| \sup_{j \in \mathbb{N}} f_{j,\epsilon} \right\|_{L^p(\mathbb{G}^n)}^p \\
& \le \sum_{j \in \mathbb{N}} \|f_{j,\epsilon}\|_{W^{1,p}(\mathbb{G}^n)}^p \\
& \le \sum_{j \in \mathbb{N}} \left(\mathrm{Cap}_p(E_j; \mathbb{G}^n) + 2^{-j}\epsilon \right),
\end{aligned}
$$

we apply Lemma 3.1.2 to deduce that ∇f_ϵ exists a. e. on $\mathbb{R}^n$ and

$$|\nabla f_\epsilon| \le \sup_{j \in \mathbb{N}} |\nabla f_{j,\epsilon}| \quad a.\,e. \text{ on } \mathbb{R}^n.$$

Notice that (3.4) also implies

$$f_\epsilon \in W^{1,p}(\mathbb{G}^n).$$

Thus,

$$f_\epsilon \in \mathcal{A}_p(\cup_{j=1}^{\infty} E_j),$$

and (3.4) again gives

$$\mathrm{Cap}_p\left(\cup_{j=1}^{\infty} E_j; \mathbb{G}^n \right) \le \|f_\epsilon\|_{W^{1,p}(\mathbb{G}^n)}^p \le \sum_{j \in \mathbb{N}} \mathrm{Cap}_p(E_j; \mathbb{G}^n) + \epsilon.$$

Letting $\epsilon \to 0$ yields (3.3). This proves (iii).

(iv) When $1 \le p < q < \infty$, using the Hölder inequality and the elementary inequality

$$(3.5) \qquad a + b \le 2^{1 - \frac{1}{\kappa}} (a^\kappa + b^\kappa)^{\frac{1}{\kappa}} \quad \forall \quad (\kappa, a, b) \in [1, \infty) \times (0, \infty) \times (0, \infty),$$

we have that if $\kappa = \frac{q}{p}$, then

$$
\begin{aligned}
(3.6) \qquad 2^{-\frac{1}{p}} \|f\|_{W^{1,p}(\mathbb{G}^n)} &= 2^{-\frac{1}{p}} \left(\left\| |\nabla f| \right\|_{L^p(\mathbb{G}^n)}^p + \|f\|_{L^p(\mathbb{G}^n)}^p \right)^{\frac{1}{p}} \\
&\le 2^{-\frac{1}{p}} \left(\left\| |\nabla f| \right\|_{L^q(\mathbb{G}^n)}^p + \|f\|_{L^q(\mathbb{G}^n)}^p \right)^{\frac{1}{p}} \\
&\le 2^{-\frac{1}{q}} \|f\|_{W^{1,q}(\mathbb{G}^n)}.
\end{aligned}
$$

For any set $E \subseteq \mathbb{R}^n$, notice that

$$\mathcal{A}_q(E) \subseteq \mathcal{A}_p(E).$$

So, we use (3.6) and (3.1) to obtain

$$2^{-\frac{1}{p}} \left(\mathrm{Cap}_p(E; \mathbb{G}^n) \right)^{\frac{1}{p}} = \inf_{f \in \mathcal{A}_p(E)} 2^{-\frac{1}{p}} \|f\|_{W^{1,p}(\mathbb{G}^n)}$$

$$\le \inf_{f \in \mathcal{A}_p(E)} 2^{-\frac{1}{q}} \|f\|_{W^{1,q}(\mathbb{G}^n)}$$

$$\le \inf_{f \in \mathcal{A}_q(E)} 2^{-\frac{1}{q}} \|f\|_{W^{1,q}(\mathbb{G}^n)}$$

$$= 2^{-\frac{1}{q}} \left(\mathrm{Cap}_q(E; \mathbb{G}^n) \right)^{\frac{1}{q}}.$$

(v) By (ii), it is trivial to find that

$$\lim_{j \to \infty} \mathrm{Cap}_p(K_j; \mathbb{G}^n) \ge \mathrm{Cap}_p \left(\cap_{j=1}^\infty K_j; \mathbb{G}^n \right).$$

It remains to prove the converse of this last inequality. To this end, we let

$$K = \cap_{j=1}^\infty K_j,$$

which is also compact. For any $\epsilon \in (0, 1)$, choose

$$f_\epsilon \in \mathcal{A}_p(K)$$

such that

$$\|f_\epsilon\|^p_{W^{1,p}(\mathbb{G}^n)} \le \mathrm{Cap}_p(K; \mathbb{G}^n) + \epsilon.$$

Notice that the compact set K is contained in the open set

$$\{x \in \mathbb{R}^n : f_\epsilon(x) \ge 1\}^\circ.$$

Since

$$K_j \searrow K \quad \text{as} \quad j \to \infty,$$

it is a basic fact that there exists some $j_0 \in \mathbb{N}$ such that

$$K_{j_0} \subseteq \{x \in \mathbb{R}^n : f_\epsilon(x) \ge 1\}^\circ.$$

Thus,

$$f_\epsilon \in \mathcal{A}_p(K_{j_0})$$

and

$$\lim_{j \to \infty} \mathrm{Cap}_p(K_j; \mathbb{G}^n) \le \mathrm{Cap}_p(K_{j_0}; \mathbb{G}^n)$$

$$\le \|f_\epsilon\|^p_{W^{1,p}(\mathbb{G}^n)}$$

$$\le \mathrm{Cap}_p(K; \mathbb{G}^n) + \epsilon.$$

Letting $\epsilon \to 0$ in the last formulae yields the desired result of (v). $\qquad \square$

Proposition 3.1.5. Let $p \in (1, \infty)$. Then for any sequence $\{E_j\}_{j=1}^{\infty}$ with

$$E_j \subseteq E_{j+1} \subseteq \mathbb{R}^n \ \forall \ j \in \mathbb{N},$$

one has

$$\lim_{j \to \infty} \mathrm{Cap}_p \left(E_j; \ \mathbb{G}^n \right) = \mathrm{Cap}_p \left(\cup_{j=1}^{\infty} E_j; \ \mathbb{G}^n \right).$$

Proof. We adopt the idea used in Costea [15, Theorem 3.1(iv)]. Let

$$E = \cup_{i=1}^{\infty} E_i.$$

By Proposition 3.1.4(ii), we obtain

$$\lim_{i \to \infty} \mathrm{Cap}_p \left(E_i; \ \mathbb{G}^n \right) \leq \mathrm{Cap}_p \left(E; \ \mathbb{G}^n \right).$$

Thus, we only need to prove the converse of the above inequality. Without loss of generality, we may as well assume

$$\lim_{i \to \infty} \mathrm{Cap}_p \left(E_i; \ \mathbb{G}^n \right) < \infty.$$

Fix $\epsilon \in (0, 1)$. For any $i \in \mathbb{N}$, choose

$$u_i \in \mathcal{A}_p(E_i)$$

such that

$$\|u_i\|_{W^{1,p}(\mathbb{G}^n)}^p \leq \mathrm{Cap}_p \left(E_i; \ \mathbb{G}^n \right) + \epsilon.$$

Since $\{E_i\}_{i=1}^{\infty}$ increases and

$$\lim_{i \to \infty} \mathrm{Cap}_p \left(E_i; \ \mathbb{G}^n \right) < \infty,$$

it follows that

$$\sup_{i \in \mathbb{N}} \|u_i\|_{W^{1,p}(\mathbb{G}^n)} < \infty.$$

Applying Proposition 1.3.1, we find a subsequence, which we denote again by $\{u_i\}_{i \in \mathbb{N}}$, and a function

$$u \in W^{1,p}(\mathbb{G}^n)$$

such that $\{(u_i, \nabla u_i)\}_{i \in \mathbb{N}}$ converges to $(u, \nabla u)$ weakly in

$$L^p(\mathbb{G}^n) \times L^p(\mathbb{G}^n; \mathbb{R}^n).$$

Upon fixing $i_0 \in \mathbb{N}$ and using Mazur's Theorem, for the sequence $\{u_i\}_{i \geq i_0}$ we find a finite convex combination of $\{u_i\}_{i \geq i_0}$, denoted by v_{i_0}, such that

(3.7) $$\|v_{i_0} - u\|_{W^{1,p}(\mathbb{G}^n)} < 2^{-i_0}.$$

Since every u_i with $i \geq i_0$ satisfies

$$E_{i_0} \subseteq E_i \subseteq \left\{x \in \mathbb{R}^n : u_i(x) \geq 1\right\}^\circ,$$

it follows that

$$E_{i_0} \subseteq \left\{x \in \mathbb{R}^n : v_{i_0}(x) \geq 1\right\}^\circ.$$

In this way, we obtain a sequence $\{v_i\}_{i \in \mathbb{N}}$ with each v_i being a finite convex combination of $\{u_k\}_{k \geq i}$ such that

$$v_i \to u \text{ strongly in } W^{1,p}(\mathbb{G}^n) \text{ as } i \to \infty,$$

and so that

$$v_i \in \mathcal{A}_p(E_i).$$

Passing to a subsequence if necessary, we may even assume that for any $i \in \mathbb{N}$,

$$(3.8) \qquad \qquad \|v_{i+1} - v_i\|_{W^{1,p}(\mathbb{G}^n)} < 2^{-i}.$$

Next, for any $j \in \mathbb{N}$, define

$$w_j = \sup_{i \geq j} v_i.$$

It is easy to verify that for all $j \in \mathbb{N}$ and $x \in \mathbb{R}^n$,

$$(3.9) \qquad \qquad |w_j(x) - v_j(x)| \leq \sum_{i=j}^{\infty} |v_{i+1}(x) - v_i(x)|.$$

and

$$(3.10) \qquad \qquad \left|\sup_{i \geq j} |\nabla v_i(x)| - |\nabla v_j(x)|\right| \leq \sum_{i=j}^{\infty} |\nabla v_{i+1}(x) - \nabla v_i(x)|.$$

By (3.8) and (3.9), we have

$$\|w_j\|_{L^p(\mathbb{G}^n)} \leq \|v_j\|_{L^p(\mathbb{G}^n)} + \sum_{i=j}^{\infty} \|v_{i+1} - v_i\|_{L^p(\mathbb{G}^n)}$$

$$(3.11) \qquad \qquad \leq \|v_j\|_{L^p(\mathbb{G}^n)} + \sum_{i=j}^{\infty} 2^{-i}$$

$$\leq \|v_j\|_{L^p(\mathbb{G}^n)} + 2^{1-j}.$$

Similarly, by (3.9) and (3.10), we obtain

$$(3.12) \qquad \left\|\sup_{i \geq j} |\nabla v_i|\right\|_{L^p(\mathbb{G}^n)} \leq \left\||\nabla v_j|\right\|_{L^p(\mathbb{G}^n)} + \sum_{i=j}^{\infty} \left\||\nabla v_{i+1} - \nabla v_i|\right\|_{L^p(\mathbb{G}^n)}$$

$$\leq \left\||\nabla v_j|\right\|_{L^p(\mathbb{G}^n)} + 2^{1-j}.$$

Notice that (3.11)-(3.12) and Lemma 3.1.2 imply that ∇w_j exists a. e. on $\mathbb{R}^n$ and

$$(3.13) \qquad |\nabla w_j(x)| \leq \sup_{i \geq j} |\nabla v_i(x)| \quad \text{a.e.} \ \ x \in \mathbb{R}^n.$$

By (3.13)-(3.12)-(3.11), we see

$$w_j \in W^{1,p}(\mathbb{G}^n).$$

Now, we calculate the $\| \cdot \|_{W^{1,p}(\mathbb{G}^n)}$ norm of w_j. To this end, we observe that the mean value theorem implies the following inequality:

$$(3.14) \quad (a+b)^p \leq a^p + p(M+1)^{p-1}b \quad \text{for} \quad (M,a,b) \in (0,\infty) \times [0,M] \times [0,1].$$

Notice that (3.7) implies

$$\max \left\{ \|v_j\|_{L^p(\mathbb{G}^n)}, \, \||\nabla v_j|\|_{L^p(\mathbb{G}^n)} \right\} \leq \|u\|_{W^{1,p}(\mathbb{G}^n)} + 1.$$

Below we will apply (3.14) with

$$M = \|u\|_{W^{1,p}(\mathbb{G}^n)} + 1.$$

Consequently, we deduce from (3.11), (3.12), and (3.13) that

$$
\begin{aligned}
\|w_j\|_{W^{1,p}(\mathbb{G}^n)}^p &\leq \|w_j\|_{L^p(\mathbb{G}^n)}^p + \left\| \sup_{i \geq j} |\nabla v_i| \right\|_{L^p(\mathbb{G}^n)}^p \\
(3.15) \qquad &\leq \left(\|v_j\|_{L^p(\mathbb{G}^n)} + 2^{1-j} \right)^p + \left(\||\nabla v_j|\|_{L^p(\mathbb{G}^n)} + 2^{1-j} \right)^p \\
&\leq \|v_j\|_{W^{1,p}(\mathbb{G}^n)}^p + C_{u,p} 2^{-j},
\end{aligned}
$$

where

$$C_{u,p} := 4p(\|u\|_{W^{1,p}(\mathbb{G}^n)} + 2)^{p-1}.$$

Recalling that

$$v_i \in \mathcal{A}_p(E_i)$$

and $\{E_i\}_i$ increases to E, we have

$$E \subseteq \left\{ x \in \mathbb{R}^n : w_j(x) \geq 1 \right\}^{\circ}.$$

Thus,

$$w_j \in \mathcal{A}_p(E).$$

Moreover, (3.15) implies

$$\text{Cap}_p (E; \, \mathbb{G}^n) \leq \|w_j\|_{W^{1,p}(\mathbb{G}^n)}^p \leq \|v_j\|_{W^{1,p}(\mathbb{G}^n)}^p + C_{u,p} 2^{-j} \quad \forall \ \ j \in \mathbb{N}.$$

According to the construction of v_j, we may assume

$$v_j = \sum_{k=j}^{N_j} \lambda_{j,k} u_k,$$

where

$$\begin{cases} j \le N_j \in \mathbb{N}; \\ \lambda_{j,k} \in [0,1]; \\ \sum_{k=j}^{N_j} \lambda_{j,k} = 1. \end{cases}$$

Consequently, by the Minkowski inequality, the Hölder inequality, and Proposition 3.1.4 (ii), we achieve

$$\|v_j\|_{W^{1,p}(\mathbb{G}^n)}^p = \|v_j\|_{L^p(\mathbb{G}^n)}^p + \||\nabla v_j|\|_{L^p(\mathbb{G}^n)}^p$$

$$\le \left(\sum_{k=j}^{N_j} \lambda_{j,k} \|u_k\|_{L^p(\mathbb{G}^n)} \right)^p + \left(\sum_{k=j}^{N_j} \lambda_{j,k} \||\nabla u_k|\|_{L^p(\mathbb{G}^n)} \right)^p$$

$$\le \sum_{k=j}^{N_j} \lambda_{j,k} \|u_k\|_{L^p(\mathbb{G}^n)}^p + \sum_{k=j}^{N_j} \lambda_{j,k} \||\nabla u_k|\|_{L^p(\mathbb{G}^n)}^p$$

$$= \sum_{k=j}^{N_j} \lambda_{j,k} \|u_k\|_{W^{1,p}(\mathbb{G}^n)}^p$$

$$\le \sum_{k=j}^{N_j} \lambda_{j,k} \left(\mathrm{Cap}_p (E_k; \mathbb{G}^n) + \epsilon \right)$$

$$\le \mathrm{Cap}_p \left(E_{N_j}; \mathbb{G}^n \right) + \epsilon.$$

We then deduce that for any $j \in \mathbb{N}$,

$$\mathrm{Cap}_p (E; \mathbb{G}^n) \le \mathrm{Cap}_p \left(E_{N_j}; \mathbb{G}^n \right) + \epsilon + C_{u,p} 2^{-j}.$$

Letting $j \to \infty$ and $\epsilon \to 0$ yields

$$\mathrm{Cap}_p (E; \mathbb{G}^n) \le \lim_{j \to \infty} \mathrm{Cap}_p \left(E_{N_j}; \mathbb{G}^n \right) = \lim_{j \to \infty} \mathrm{Cap}_p (E_j; \mathbb{G}^n),$$

thereby completing the argument. $\qquad\qquad\qquad\qquad\qquad\qquad\qquad\square$

Remark 3.1.6. The limiting case $p \to 1$ of Proposition 3.1.5 will be presented through Theorem 6.2.1(v) of Chapter 6 for the Gaussian BV-capacity.

3.2 Alternative of Gaussian p-Capacity for $1 \leq p < \infty$

Definition 3.2.1. Let $p \in [1, \infty)$ and $K \subseteq \mathbb{R}^n$ be a compact set. Define

$$\mathcal{A}(K) := \{f \in C_c^1(\mathbb{R}^n) : \ f \geq 1 \text{ on } K\}$$

and

(3.16) $\mathrm{Cap}_{0,p}(K; \mathbb{G}^n) := \inf \left\{ \|f\|_{W^{1,p}(\mathbb{G}^n)}^p : \ f \in \mathcal{A}(K) \right\}.$

If $O \subseteq \mathbb{R}^n$ is an open set, then

(3.17) $\mathrm{Cap}_{0,p}(O; \mathbb{G}^n) := \sup \left\{ \mathrm{Cap}_{0,p}(K; \mathbb{G}^n) : \text{ compact } K \subseteq O \right\}.$

Remark 3.2.2. As in Remark 3.1.3(ii)-(iii), by

$$\|f\|_{W^{1,p}(\mathbb{G}^n)} \leq \|f\|_{W^{1,p}(\mathbb{G}^n)} \quad \forall \ f \in C_c^1(\mathbb{R}^n),$$

we also have

$$\mathrm{Cap}_{0,p}(K; \mathbb{G}^n) = \inf \left\{ \|f\|_{W^{1,p}(\mathbb{G}^n)}^p : \ 0 \leq f \in \mathcal{A}(K) \right\}$$
$$= \inf \left\{ \|f\|_{W^{1,p}(\mathbb{G}^n)}^p : \ 0 \leq f \leq 1, f \in \mathcal{A}(K) \right\}.$$

Lemma 3.2.3. Let $p \in [1, \infty)$. Then one has the following properties.

(i) For compact sets K_1 and K_2 satisfying that $K_1 \subseteq K_2 \subseteq \mathbb{R}^n$,

$$\mathrm{Cap}_{0,p}(K_1; \mathbb{G}^n) \leq \mathrm{Cap}_{0,p}(K_2; \mathbb{G}^n).$$

(ii) For compact set K and open set O satisfying $O \subseteq K \subseteq \mathbb{R}^n$,

$$\mathrm{Cap}_{0,p}(O; \mathbb{G}^n) \leq \mathrm{Cap}_{0,p}(K; \mathbb{G}^n).$$

(iii) For open sets O_1 and O_2 satisfying $O_1 \subseteq O_2 \subseteq \mathbb{R}^n$,

$$\mathrm{Cap}_{0,p}(O_1; \mathbb{G}^n) \leq \mathrm{Cap}_{0,p}(O_2; \mathbb{G}^n).$$

Proof. First of all, if K_1 and K_2 are compact subsets of $\mathbb{R}^n$ satisfying $K_1 \subseteq K_2$, then

$$\mathcal{A}(K_2) \subseteq \mathcal{A}(K_1),$$

and hence

$$\mathrm{Cap}_p(K_1; \mathbb{G}^n) \leq \mathrm{Cap}_p(K_2; \mathbb{G}^n),$$

which proves (i).

Next, (ii) follows from (i) and (3.17).

Finally, (iii) follows directly from (ii) and (3.17).   □

Lemma 3.2.4. Let $p \in [1, \infty)$ and K be a compact subset of $\mathbb{R}^n$. Then

$$(3.18) \qquad \text{Cap}_{0,p}(K; \mathbb{G}^n) = \inf \left\{ \text{Cap}_{0,p}(O; \mathbb{G}^n) : \text{ open } O \supseteq K \right\}.$$

Proof. By (3.17), we see that for any open set $O \supseteq K$,

$$\text{Cap}_{0,p}(K; \mathbb{G}^n) \le \text{Cap}_{0,p}(O; \mathbb{G}^n).$$

Taking the infimum over all such sets O yields

$$\text{Cap}_{0,p}(K; \mathbb{G}^n) \le \inf \left\{ \text{Cap}_{0,p}(O; \mathbb{G}^n) : \text{ open } O \supseteq K \right\}.$$

To prove the converse of this inequality, it suffices to verify that, for any $\epsilon \in (0, \infty)$, there exists an open set $O \supseteq K$ such that

$$(3.19) \qquad \text{Cap}_{0,p}(O; \mathbb{G}^n) \le (1 + \epsilon)^p \left(\text{Cap}_{0,p}(K; \mathbb{G}^n) + \epsilon \right).$$

By (3.16), there exists a function

$$f \in C_c^1(\mathbb{R}^n) \text{ such that } f \ge 1 \text{ on } K$$

and

$$\|f\|_{W^{1,p}(\mathbb{G}^n)}^p < \text{Cap}_{0,p}(K; \mathbb{G}^n) + \epsilon.$$

Define

$$\begin{cases} f_\epsilon := (1 + \epsilon)f; \\ K_\epsilon = \left\{ x \in \mathbb{R}^n : f_\epsilon(x) \ge 1 \right\}. \end{cases}$$

Since

$$f_\epsilon \in C_c^1(\mathbb{R}^n) \text{ and } f_\epsilon > 1 \text{ on } K,$$

it follows that K_ϵ is a compact set with

$$K \subseteq K_\epsilon^\circ \subseteq K_\epsilon.$$

This, along with Lemma 3.2.3(ii), implies

$$\text{Cap}_{0,p}(K_\epsilon^\circ; \mathbb{G}^n) \le \text{Cap}_{0,p}(K_\epsilon; \mathbb{G}^n).$$

Upon noticing

$$f_\epsilon \in \mathcal{A}(K_\epsilon),$$

we have

$$\text{Cap}_{0,p}(K_\epsilon; \mathbb{G}^n) \le \|f_\epsilon\|_{W^{1,p}(\mathbb{G}^n)}^p < (1 + \epsilon)^p \left(\text{Cap}_{0,p}(K; \mathbb{G}^n) + \epsilon \right).$$

Via combining the last two inequalities, we find an open set

$$O = K_\epsilon^\circ \supseteq K$$

such that (3.19) holds, thereby completing the argument for (3.18). $\qquad \square$

Due to Lemma 3.2.4, we can extend the definition of $\mathrm{Cap}_{0,p}(\cdot; \mathbb{G}^n)$ from a compact set to any set.

Definition 3.2.5. Let $p \in [1, \infty)$ and E be an arbitrary subset of $\mathbb{R}^n$. Define

$$(3.20) \qquad \mathrm{Cap}_{0,p}(E; \mathbb{G}^n) := \inf \left\{ \mathrm{Cap}_{0,p}(O; \mathbb{G}^n) : \text{ open } O \supseteq E \right\}.$$

Applying Proposition 1.1.3, we give the following equivalent characterization of the Gaussian p-capacity.

Theorem 3.2.6. If $p \in (1, \infty)$ and E is a subset of $\mathbb{R}^n$, then

$$(3.21) \qquad \mathrm{Cap}_{0,p}(E; \mathbb{G}^n) = \mathrm{Cap}_p(E; \mathbb{G}^n)$$

Also, if K is a compact subset of $\mathbb{R}^n$, then

$$(3.22) \qquad \mathrm{Cap}_{0,1}(K; \mathbb{G}^n) = \mathrm{Cap}_1(K; \mathbb{G}^n).$$

Proof. We will prove (3.21) and (3.22) according to the following three steps.

Step 1. $E = K$ is compact. For any $\epsilon > 0$, by (3.16) and the definition of $\mathcal{A}(K)$, there exists

$$f \in C_c^1(\mathbb{R}^n) \text{ such that } f \geq 1 \text{ on } K$$

and

$$\mathrm{Cap}_{0,p}(K; \mathbb{G}^n) \leq \|f\|^p_{W^{1,p}(\mathbb{G}^n)} \leq \mathrm{Cap}_{0,p}(K; \mathbb{G}^n) + \epsilon.$$

For any $\lambda \in (0, \infty)$ define

$$f_\lambda := (1 + \lambda)f.$$

Since f is continuous, it follows that

$$\begin{cases} K \subseteq \{x \in \mathbb{R}^n : f_\lambda(x) > 1\} \subseteq \{x \in \mathbb{R}^n : f_\lambda(x) \geq 1\}^\circ; \\ f_\lambda \in \mathcal{A}_p(K). \end{cases}$$

Accordingly,

$$\begin{aligned} \mathrm{Cap}_p(K; \mathbb{G}^n) &\leq \|f_\lambda\|^p_{W^{1,p}(\mathbb{G}^n)} \\ &= (1 + \lambda)^p \|f\|^p_{W^{1,p}(\mathbb{G}^n)} \\ &\leq (1 + \lambda)^p \left(\mathrm{Cap}_{0,p}(K; \mathbb{G}^n) + \epsilon \right). \end{aligned}$$

In the above inequality, letting $\epsilon \to 0$ and $\lambda \to 0$ yields

$$(3.23) \qquad \mathrm{Cap}_p(K; \mathbb{G}^n) \leq \mathrm{Cap}_{0,p}(K; \mathbb{G}^n).$$

Starting from the definition of $\mathrm{Cap}_p(K; \mathbb{G}^n)$, we see that, for any $\epsilon \in (0, \infty)$, there exists a function

$$f \in \mathcal{A}_p(K) \text{ such that } \mathrm{Cap}_p(K; \mathbb{G}^n) \leq \|f\|^p_{W^{1,p}(\mathbb{G}^n)} \leq \mathrm{Cap}_p(K; \mathbb{G}^n) + \epsilon.$$

It follows from the definition of $\mathcal{A}_p(K)$ that

$$f \in W^{1,p}(\mathbb{G}^n) \quad \& \quad K \subseteq \{x \in \mathbb{R}^n : f(x) \geq 1\}^\circ.$$

By Proposition 1.1.3, there exists a sequence of functions

$$\{f_j\}_{j \in \mathbb{N}} \subseteq C_c^1(\mathbb{R}^n)$$

such that

$$\lim_{j \to \infty} \|f_j - f\|_{W^{1,p}(\mathbb{G}^n)} = 0$$

and each f_j satisfies

$$f_j \geq 1 \quad \text{on } K.$$

Therefore, we obtain:

$$\begin{cases} f_j \in \mathcal{A}(K); \\ \mathrm{Cap}_{0,p}(K; \mathbb{G}^n) \leq \lim_{j \to \infty} \|f_j\|_{W^{1,p}(\mathbb{G}^n)}^p = \|f\|_{W^{1,p}(\mathbb{G}^n)}^p \leq \mathrm{Cap}_p(K; \mathbb{G}^n) + \epsilon. \end{cases}$$

Letting $\epsilon \to 0$ in the above inequality implies

$$(3.24) \qquad\qquad \mathrm{Cap}_{0,p}(K; \mathbb{G}^n) \leq \mathrm{Cap}_p(K; \mathbb{G}^n).$$

Combining (3.23) and (3.24) implies that (3.21) and (3.22) hold when $E = K$ is a compact subset of $\mathbb{R}^n$.

Step 2. $E = O$ *is open.* For each compact set $K \subseteq O$, it is easy to verify

$$\mathcal{A}_p(O) \subseteq \mathcal{A}_p(K).$$

By this, (3.17) and *Step 1*, we get that if $p \in [1, \infty)$ then

$$\begin{aligned} \mathrm{Cap}_{0,p}(O; \mathbb{G}^n) &= \sup_{\text{compact } K \subseteq O} \mathrm{Cap}_{0,p}(K; \mathbb{G}^n) \\ &= \sup_{\text{compact } K \subseteq O} \mathrm{Cap}_p(K; \mathbb{G}^n) \\ &= \sup_{\text{compact } K \subseteq O} \inf_{f \in \mathcal{A}_p(K)} \|f\|_{W^{1,p}(\mathbb{G}^n)}^p \\ &\leq \inf_{f \in \mathcal{A}_p(O)} \|f\|_{W^{1,p}(\mathbb{G}^n)}^p \\ &= \mathrm{Cap}_p(O; \mathbb{G}^n). \end{aligned}$$

It remains to prove

$$\mathrm{Cap}_p(O; \mathbb{G}^n) \leq \mathrm{Cap}_{0,p}(O; \mathbb{G}^n) \quad \text{as} \quad p \in (1, \infty).$$

Since O is an open set of $\mathbb{R}^n$, there exists an increasing sequence of compact sets $\{K_j\}_{j \in \mathbb{N}}$ such that

$$\cup_{j=1}^{\infty} K_j = O.$$

Applying (vi) of Proposition 3.1.4, the already-proved result in *Step 1* and (3.17), we conclude

$$
\begin{aligned}
\mathrm{Cap}_p(O;\,\mathbb{G}^n) &= \lim_{j\to\infty} \mathrm{Cap}_p(K_j;\,\mathbb{G}^n) \\
&= \lim_{j\to\infty} \mathrm{Cap}_{0,p}(K_j;\,\mathbb{G}^n) \\
&\leq \sup_{\text{compact } K\subseteq O} \mathrm{Cap}_{0,p}(K;\,\mathbb{G}^n) \\
&= \mathrm{Cap}_{0,p}(O;\,\mathbb{G}^n).
\end{aligned}
$$

Hence, we obtain (3.21) when $E = O$ is open.

Step 3. E is an arbitrary set. According to (3.20), for any $\epsilon > 0$ there exists an open set $O \supseteq E$ such that

$$
\mathrm{Cap}_{0,p}(O;\,\mathbb{G}^n) \leq \mathrm{Cap}_{0,p}(E;\,\mathbb{G}^n) + \epsilon.
$$

By the definition of $\mathrm{Cap}_p(O;\,\mathbb{G}^n)$ and *Step 2*, we can find an $f \in \mathcal{A}_p(O)$ such that

$$
\|f\|^p_{W^{1,p}(\mathbb{G}^n)} \leq \mathrm{Cap}_p(O;\,\mathbb{G}^n) + \epsilon = \mathrm{Cap}_{0,p}(O;\,\mathbb{G}^n) + \epsilon.
$$

Since

$$
O \supseteq E \Rightarrow \mathcal{A}_p(O) \subseteq \mathcal{A}_p(E),
$$

the above f also belongs to $\mathcal{A}_p(E)$, which, along with the last two formulae, yields

$$
\mathrm{Cap}_p(E;\,\mathbb{G}^n) \leq \|f\|^p_{W^{1,p}(\mathbb{G}^n)} \leq \mathrm{Cap}_{0,p}(O;\,\mathbb{G}^n) + \epsilon \leq \mathrm{Cap}_{0,p}(E;\,\mathbb{G}^n) + 2\epsilon.
$$

Then letting $\epsilon \to 0$ gives

$$
\mathrm{Cap}_p(E;\,\mathbb{G}^n) \leq \mathrm{Cap}_{0,p}(E;\,\mathbb{G}^n).
$$

Now we prove the converse of this last inequality. For any $\epsilon > 0$, by the definition of $\mathrm{Cap}_p(E;\,\mathbb{G}^n)$, there exists

$$
f \in \mathcal{A}_p(E) \text{ such that } \|f\|^p_{W^{1,p}(\mathbb{G}^n)} \leq \mathrm{Cap}_p(E;\,\mathbb{G}^n) + \epsilon.
$$

The definition of $\mathcal{A}_p(E)$ implies

$$
f \in W^{1,p}(\mathbb{G}^n) \quad \& \quad E \subseteq \{x \in \mathbb{R}^n : f(x) \geq 1\}^\circ =: U.
$$

Observe that U is open with

$$
\mathcal{A}_p(U) \subseteq \mathcal{A}_p(E) \quad \& \quad f \in \mathcal{A}_p(U).
$$

So, from the definition of $\mathrm{Cap}_{0,p}(E;\,\mathbb{G}^n)$ and *Step 2*, it follows that

$$
\mathrm{Cap}_{0,p}(E;\,\mathbb{G}^n) \leq \mathrm{Cap}_{0,p}(U;\,\mathbb{G}^n)
$$

$$= \mathrm{Cap}_p(U;\ \mathbb{G}^n)$$

$$\le \|f\|^p_{W^{1,p}(\mathbb{G}^n)}$$

$$\le \mathrm{Cap}_p(E;\ \mathbb{G}^n) + \epsilon.$$

Letting $\epsilon \to 0$ yields

$$\mathrm{Cap}_{0,p}(E;\ \mathbb{G}^n) \le \mathrm{Cap}_p(E;\ \mathbb{G}^n).$$

Thus, we derive that (3.21) holds for any general set E. □

Note that Definition 3.2.5 and Theorem 3.2.6 reveal that $\mathrm{Cap}_{1<p<\infty}(\cdot;\mathbb{G}^n)$ possesses the outer regularity. Next, we prove the inner regularity for all Borel sets.

Proposition 3.2.7. Under $p \in (1,\infty)$, any Borel subset E of $\mathbb{R}^n$ satisfies

(3.25) $$\mathrm{Cap}_p(E;\ \mathbb{G}^n) = \sup\left\{\mathrm{Cap}_p(K;\ \mathbb{G}^n):\ \text{compact } K \subseteq E\right\}.$$

Proof. For $p \in (1,\infty)$, since $\mathrm{Cap}_p(\cdot;\mathbb{G}^n)$ satisfies (i)-(ii)-(iii) and (v) of Proposition 3.1.4 and Proposition 3.1.5, it follows from Adams-Hedberg's book [2, Theorem 2.3.11] that (3.25) holds for all Borel subsets of $\mathbb{R}^n$. □

Chapter 4
Restriction of Gaussian Sobolev p-Space

Based on Chapters 1–3, in this chapter we settle the restriction/trace question asked in the preface of this monograph.

4.1 Gaussian p-Capacitary-Strong-Type Inequality

Lemma 4.1.1. Let $1 \leq p < \infty$ and $f \in W^{1,p}(\mathbb{G}^n)$ be continuous. For any $t \in (0, \infty)$ set

$$E_t(f) := \{x \in \mathbb{R}^n : |f(x)| > t\}.$$

Then

(4.1)
$$\int_0^\infty \operatorname{Cap}_p(E_t(f); \mathbb{G}^n) \, dt^p \lesssim \|f\|_{W^{1,p}(\mathbb{G}^n)}^p.$$

Proof. Let

$$f \in W^{1,p}(\mathbb{G}^n)$$

be a nonzero continuous function. In what follows, for any $t \in (0, \infty)$, we simply write $E_t(f)$ as E_t. By Proposition 3.1.4(ii),

(4.2)
$$\int_0^\infty \operatorname{Cap}_p(E_t; \mathbb{G}^n) \, dt^p = \sum_{k \in \mathbb{Z}} \int_{2^k}^{2^{k+1}} \operatorname{Cap}_p(E_t; \mathbb{G}^n) \, dt^p$$
$$\leq (2^p - 1) \sum_{k \in \mathbb{Z}} 2^{kp} \operatorname{Cap}_p(E_{2^k}; \mathbb{G}^n).$$

As in [41, p. 155, Remark 1], we choose a function $\tau : \mathbb{R} \to \mathbb{R}$ such that

$$\begin{cases} \tau \in C^1(\mathbb{R}) & \text{is even;} \\ 0 \leq \tau(t) \leq 1 & \forall\, t \geq 0; \\ \tau(t) = 0 & \forall\, t \in [0, 2^{-1}]; \\ \tau(t) = 1 & \forall\, t \geq 1; \\ 0 \leq \tau'(t) \leq 3 & \forall\, t \geq 0. \end{cases}$$

© Springer Nature Switzerland AG 2018
L. Liu et al., *Gaussian Capacity Analysis*, Lecture Notes in Mathematics 2225,
https://doi.org/10.1007/978-3-319-95040-2_4

Indeed, on the interval $[2^{-1}, 1]$, we can define τ by smoothing the line passing over the points $(2^{-1}, 0)$ and $(1, 1)$ so that

$$0 \leq \tau'(t) \leq 2 + \epsilon$$

for some small $\epsilon \in (0, 1)$. For all $k \in \mathbb{Z}$ and $x \in \mathbb{R}^n$, we define

$$f_k(x) := \tau\left(\frac{f(x)}{2^k}\right)$$

which is also a continuous function owing to

$$f \in C^0(\mathbb{R}^n).$$

For all $k \in \mathbb{Z}$, observe that

$$f_k(x) = \tau\left(\frac{f(x)}{2^k}\right) = \tau\left(\frac{|f(x)|}{2^k}\right) = \begin{cases} 0 & \forall \ x \notin E_{2^{k-1}}; \\ 1 & \forall \ x \in E_{2^k}. \end{cases}$$

and

$$0 \leq f_k(x) \leq 1 \quad \forall \quad x \in E_{2^{k-1}} \setminus E_{2^k}.$$

Thus

$$\int_{\mathbb{R}^n} |f_k|^p \, dV_\gamma < V_\gamma(\mathbb{R}^n) = 1$$

and

$$\int_{\mathbb{R}^n} |\nabla f_k|^p \, dV_\gamma = \int_{\mathbb{R}^n} \left| \tau'\left(\frac{f(x)}{2^k}\right) \frac{\nabla f(x)}{2^k} \right|^p \, dV_\gamma \leq 3^p \int_{E_{2^{k-1}} \setminus E_{2^k}} |\nabla f(x)|^p 2^{-kp} \, dV_\gamma,$$

which implies

$$f_k \in W^{1,p}(\mathbb{G}^n).$$

Also, since $f_k = 1$ on the open set E_{2^k}, it is easy to verify

$$E_{2^k} \subseteq \{x \in \mathbb{R}^n : f_k(x) \geq 1\}^\circ.$$

Hence,

$$f_k \in \mathcal{A}_p(E_{2^k})$$

and

$$
\begin{aligned}
\text{Cap}_p(E_{2^k}; \mathbb{G}^n) &\leq \|f_k\|_{W^{1,p}(\mathbb{G}^n)}^p \\
&\leq \int_{\mathbb{R}^n} |\nabla f_k|^p \, dV_\gamma + \int_{\mathbb{R}^n} |f_k|^p \, dV_\gamma \\
&\leq \int_{E_{2^{k-1}} \setminus E_{2^k}} |\nabla f(x)|^p 2^{-kp} \, dV_\gamma(x) + V_\gamma(E_{2^{k-1}}).
\end{aligned}
$$

Combining this with (4.2) yields

$$
\begin{aligned}
(4.3) \quad & \int_0^\infty \mathrm{Cap}_p(E_t; \mathbb{G}^n)\, dt^p \\
& \lesssim \sum_{k \in \mathbb{Z}} \int_{E_{2^{k-1}} \setminus E_{2^k}} |\nabla f(x)|^p \, dV_\gamma + \sum_{k \in \mathbb{Z}} 2^{kp} V_\gamma(E_{2^{k-1}}) \\
& \lesssim \|\nabla f\|^p_{L^p(\mathbb{G}^n)} + \sum_{k \in \mathbb{Z}} 2^{kp} V_\gamma(E_{2^{k-1}}).
\end{aligned}
$$

By the Hölder inequality, we have

$$
\begin{aligned}
\sum_{k \in \mathbb{Z}} 2^{kp} V_\gamma(E_{2^{k-1}}) &= \sum_{k \in \mathbb{Z}} 2^{kp} \sum_{i=k-1}^{\infty} V_\gamma(E_{2^i} \setminus E_{2^{i+1}}) \\
&= \frac{2^p}{1 - 2^{-p}} \sum_{i \in \mathbb{Z}} 2^{ip} V_\gamma(E_{2^i} \setminus E_{2^{i+1}}) \\
&\lesssim \|f\|^p_{L^p(\mathbb{G}^n)}.
\end{aligned}
$$

From this and (4.3), it follows that (4.1) holds. $\square$

4.2 Trace Inequality for $W^{1,p}(\mathbb{G}^n)$ Under $1 \leq p \leq q < \infty$

Now we use Lemma 4.1.1 to establish the first restriction/trace result for $W^{1,p}(\mathbb{G}^n)$.

Theorem 4.2.1. Let $1 \leq p \leq q < \infty$ and μ be a nonnegative Radon measure on $\mathbb{R}^n$. Then the following two assertions are equivalent.

(i) There exists a positive constant C_1 such that for all compact sets $K \subseteq \mathbb{R}^n$,

$$
\mu(K) \leq C_1 \big(\mathrm{Cap}_p(K; \mathbb{G}^n)\big)^{\frac{q}{p}}.
$$

(ii) There exists a positive constant C_2 such that

$$
(4.4) \quad \left(\int_{\mathbb{R}^n} |f|^q \, d\mu \right)^{\frac{1}{q}} \leq C_2 \|f\|_{W^{1,p}(\mathbb{G}^n)} \quad \forall \ f \in C^0(\mathbb{R}^n) \cap W^{1,p}(\mathbb{G}^n).
$$

Moreover, $C_1 \simeq C_2^q$ with the implicit constants depending only on p and q.

Proof. Assuming (i) we prove (ii). Fix

$$
f \in C^0(\mathbb{R}^n) \cap W^{1,p}(\mathbb{G}^n).
$$

For $t \in (0, \infty)$, let

$$E_t := \{x \in \mathbb{R}^n : |f(x)| > t\},$$

which is open. Consider the set E_{2^k} with $k \in \mathbb{Z}$. Since μ is a nonnegative Radon measure on $\mathbb{R}^n$, we can find a compact set $K_k \subseteq E_{2^k}$ such that

$$\mu(E_{2^k}) \leq 2\mu(K_k).$$

By this last inequality and Proposition 3.1.4(ii), we have

$$
\begin{aligned}
\int_{\mathbb{R}^n} |f|^q \, d\mu &= \sum_{k \in \mathbb{Z}} \int_{2^k}^{2^{k+1}} \mu(E_t) \, dt^q \\
&\leq (2^q - 1) \sum_{k \in \mathbb{Z}} 2^{kq} \mu(E_{2^k}) \\
&\leq 2^{q+1} \sum_{k \in \mathbb{Z}} 2^{kq} \mu(K_k) \\
&\leq C_1 2^{q+1} \sum_{k \in \mathbb{Z}} 2^{kq} \left(\mathrm{Cap}_p(K_k; \mathbb{G}^n) \right)^{\frac{q}{p}} \\
&\leq C_1 2^{q+1} \sum_{k \in \mathbb{Z}} 2^{kq} \left(\mathrm{Cap}_p(E_{2^k}; \mathbb{G}^n) \right)^{\frac{q}{p}}
\end{aligned}
$$

(4.5)

We shall use the following inequality: for any nonnegative sequence $\{a_j\}_{j \in \mathbb{Z}}$,

$$\left(\sum_{j \in \mathbb{Z}} a_j \right)^{\kappa} \leq \sum_{j \in \mathbb{Z}} a_j^{\kappa} \quad \forall \quad \kappa \in (0, 1].$$

This estimation, along with the fact $p \leq q$, further yields

$$(4.6) \qquad \sum_{k \in \mathbb{Z}} 2^{kq} \left(\mathrm{Cap}_p(E_{2^k}; \mathbb{G}^n) \right)^{\frac{q}{p}} \leq \left(\sum_{k \in \mathbb{Z}} 2^{kp} \, \mathrm{Cap}_p(E_{2^k}; \mathbb{G}^n) \right)^{\frac{q}{p}}$$

Again, using Proposition 3.1.4(ii) we obtain

$$
\begin{aligned}
\sum_{k \in \mathbb{Z}} 2^{kp} \mathrm{Cap}_p(E_{2^k}; \mathbb{G}^n) &= (1 - 2^{-p})^{-1} \sum_{k \in \mathbb{Z}} \int_{2^{k-1}}^{2^k} \mathrm{Cap}_p(E_{2^k}; \mathbb{G}^p) \, dt^p \\
&\leq (1 - 2^{-p})^{-1} \sum_{k \in \mathbb{Z}} \int_{2^{k-1}}^{2^k} \mathrm{Cap}_p(E_t; \mathbb{G}^n) \, dt^p \\
&= (1 - 2^{-p})^{-1} \int_0^{\infty} \mathrm{Cap}_p(E_t; \mathbb{G}^n) \, dt^p.
\end{aligned}
$$

(4.7)

Combining (4.5), (4.6), (4.7), and Theorem 3.2.6 yields

$$\int_{\mathbb{R}^n} |f|^q \, d\mu \leq C_1 2^{q+1} (1 - 2^{-p})^{-\frac{q}{p}} \left(\int_0^{\infty} \mathrm{Cap}_p(E_t; \mathbb{G}^n) \, dt^p \right)^{\frac{q}{p}},$$

which, together with Lemma 4.1.1, yields (4.4). Thus, (ii) holds and

$$C_2^q \leq 2^{q+1}(1 - 2^{-p})^{-\frac{q}{p}} C_{\text{Lemma 4.1.1}} C_1,$$

where $C_{\text{Lemma 4.1.1}}$ is the positive constant determined in Lemma 4.1.1.

Now we show that (ii) implies (i). By Theorem 3.2.6, it suffices to prove that (i) holds for $\text{Cap}_{0,p}(\cdot; \mathbb{G}^n)$. Let K be a compact subset of $\mathbb{R}^n$. For any $f \in \mathcal{A}(K)$ we have

$$f \in C_c^1(\mathbb{R}^n) \quad \& \quad f|_K \geq 1.$$

Clearly, (4.4) holds for such f. From this and the fact that μ is nonnegative, it follows that

$$(\mu(K))^{\frac{1}{q}} \leq \left(\int_{\mathbb{R}^n} |f|^q \, d\mu \right)^{\frac{1}{q}} \leq C_2 \|f\|_{W^{1,p}(\mathbb{G}^n)}.$$

Taking infimum over all $f \in \mathcal{A}(K)$ yields

$$(\mu(K))^{\frac{1}{q}} \leq C_2 \left(\text{Cap}_{0,p}(E; \mathbb{G}^n) \right)^{\frac{1}{p}}.$$

Thus (i) holds with $C_1 \leq C_2^q$. $\square$

4.3 Trace Inequality for $W^{1,p}(\mathbb{G}^n)$ Under $0 < q < p < \infty$

The second restriction/trace result for $W^{1,p}(\mathbb{G}^n)$ is presented below.

Theorem 4.3.1. Let $p \in [1, \infty)$, $0 < q < p < \infty$, and μ be a nonnegative Radon measure. Then the following two conditions are equivalent.

(i) The function

$$(0, \infty) \ni t \mapsto h_{\mu,p}(t) := \inf \left\{ \text{Cap}_p(K; \mathbb{G}^n) : \mathbb{R}^n \supseteq K \text{ is compact with } \mu(K) \geq t \right\}$$

satisfies

(4.8) $$\|h_{\mu,p}\| := \left(\int_0^\infty \frac{ds^{\frac{p}{p-q}}}{(h_{\mu,p}(s))^{\frac{q}{p-q}}} \right)^{\frac{p-q}{p}} < \infty.$$

(ii) There exists a positive constant C such that

(4.9) $$\left(\int_{\mathbb{R}^n} |f|^q \, d\mu \right)^{\frac{1}{q}} \leq C \|f\|_{W^{1,p}(\mathbb{G}^n)} \ \forall \ f \in C^0(\mathbb{R}^n) \cap W^{1,p}(\mathbb{G}^n).$$

Moreover,

$$\|h_{\mu,p}\| \simeq C^q$$

whose implicit constants depend only on p and q.

Proof. (i) $\Rightarrow$ (ii) Fix

$$f \in C^0(\mathbb{R}^n) \cap W^{1,p}(\mathbb{G}^n).$$

For $t \in (0, \infty)$, let

$$E_t := \{x \in \mathbb{R}^n : |f(x)| > t\}.$$

Then

$$\int_{\mathbb{R}^n} |f|^q \, d\mu \le (2^q - 1) \sum_{k \in \mathbb{Z}} 2^{kq} \mu(E_{2^k})$$

$$(4.10) \qquad = (2^q - 1) \sum_{k \in \mathbb{Z}} \sum_{i=k}^{\infty} 2^{kq} \mu(E_{2^i} \setminus E_{2^{i+1}})$$

$$\lesssim \sum_{i \in \mathbb{Z}} 2^{iq} \big(\mu(E_{2^i}) - \mu(E_{2^{i+1}}) \big).$$

Using $q < p$ and the Hölder inequality, we get

$$\sum_{i \in \mathbb{Z}} 2^{iq} \big(\mu(E_{2^i}) - \mu(E_{2^{i+1}}) \big)$$

$$(4.11) \qquad = \sum_{i \in \mathbb{Z}} \left(2^{ip} \mathrm{Cap}_p(E_{2^i}; \mathbb{G}^n) \right)^{\frac{q}{p}} \frac{\mu(E_{2^i}) - \mu(E_{2^{i+1}})}{(\mathrm{Cap}_p(E_{2^i}; \mathbb{G}^n))^{\frac{q}{p}}}$$

$$\le \left(\sum_{i \in \mathbb{Z}} 2^{ip} \mathrm{Cap}_p(E_{2^i}; \mathbb{G}^n) \right)^{\frac{q}{p}} \left(\sum_{i \in \mathbb{Z}} \left(\frac{\mu(E_{2^i}) - \mu(E_{2^{i+1}})}{(\mathrm{Cap}_p(E_{2^i}; \mathbb{G}^n))^{\frac{q}{p}}} \right)^{(\frac{p}{q})'} \right)^{\frac{p-q}{p}}.$$

It follows from Proposition 3.1.4(ii) and Lemma 4.1.1 that

$$\sum_{i \in \mathbb{Z}} 2^{ip} \mathrm{Cap}_p(E_{2^i}; \mathbb{G}^n) = (1 - 2^{-p})^{-1} \sum_{i \in \mathbb{Z}} \int_{2^{i-1}}^{2^i} \mathrm{Cap}_p(E_{2^i}; \mathbb{G}^n) \, dt^p$$

$$\le (1 - 2^{-p})^{-1} \sum_{i \in \mathbb{Z}} \int_{2^{i-1}}^{2^i} \mathrm{Cap}_p(E_t; \mathbb{G}^n) \, dt^p$$

$$= (1 - 2^{-p})^{-1} \int_0^{\infty} \mathrm{Cap}_p(E_t; \mathbb{G}^n) \, dt^p$$

$$\lesssim \|f\|_{W^{1,p}(\mathbb{G}^n)}^p,$$

so that

$$(4.12) \qquad \left(\sum_{i \in \mathbb{Z}} 2^{ip} \mathrm{Cap}_p(E_{2^i}; \mathbb{G}^n) \right)^{\frac{q}{p}} \lesssim \|f\|_{W^{1,p}(\mathbb{G}^n)}^q.$$

Now we consider

$$\left(\sum_{i \in \mathbb{Z}} \left(\frac{\mu(E_{2^i}) - \mu(E_{2^{i+1}})}{(\mathrm{Cap}_p(E_{2^i}; \mathbb{G}^n))^{\frac{q}{p}}} \right)^{\left(\frac{p}{q}\right)'} \right)^{\frac{p-q}{p}}$$

in (4.11). Since

$$f \in C^0(\mathbb{R}^n),$$

it follows that every E_{2^i} is open. As μ is a Radon measure, there exists a compact set $K_i \subseteq E_{2^i}$ such that

$$2^{-1}\mu(E_{2^i}) \le \mu(K_i) \le \mu(E_{2^i}).$$

According to the definition of the function $h_{\mu,p}$, we see that

$$h_{\mu,p}\left(2^{-1}\mu(E_{2^i})\right) \le \mathrm{Cap}_p(K_i; \mathbb{G}^n) \le \mathrm{Cap}_p(E_{2^i}; \mathbb{G}^n),$$

where the second inequality holds because of (3.17). From this,

$$\left(\frac{p}{q}\right)' = \frac{p}{p-q} > 1,$$

the monotone increasing property of $h_{\mu,p}$, and (4.8), it follows that

$$
\begin{aligned}
\sum_{i \in \mathbb{Z}} \left(\frac{\mu(E_{2^i}) - \mu(E_{2^{i+1}})}{(\mathrm{Cap}_p(E_{2^i}; \mathbb{G}^n))^{\frac{q}{p}}} \right)^{\left(\frac{p}{q}\right)'} &\le \sum_{i \in \mathbb{Z}} \frac{(\mu(E_{2^i}) - \mu(E_{2^{i+1}}))^{\frac{p}{p-q}}}{\left(h_{\mu,p}\left(2^{-1}\mu(E_{2^i})\right)\right)^{\frac{q}{p-q}}} \\
&\le \sum_{i \in \mathbb{Z}} \frac{(\mu(E_{2^i}))^{\frac{p}{p-q}} - (\mu(E_{2^{i+1}}))^{\frac{p}{p-q}}}{\left(h_{\mu,p}\left(2^{-1}\mu(E_{2^i})\right)\right)^{\frac{q}{p-q}}} \\
\text{(4.13)} \qquad &= 2^{\frac{p}{p-q}} \sum_{i \in \mathbb{Z}} \int_{2^{-1}\mu(E_{2^{i+1}})}^{2^{-1}\mu(E_{2^i})} \frac{ds^{\frac{p}{p-q}}}{\left(h_{\mu,p}\left(2^{-1}\mu(E_{2^i})\right)\right)^{\frac{q}{p-q}}} \\
&\le 2^{\frac{p}{p-q}} \int_0^\infty \frac{ds^{\frac{p}{p-q}}}{\left(h_{\mu,p}(s)\right)^{\frac{q}{p-q}}} \\
&= 2^{\frac{p}{p-q}} \|h_{\mu,p}\|^{\frac{p}{p-q}}.
\end{aligned}
$$

Inserting the estimates of (4.12) and (4.13) into (4.11) yields

$$\sum_{i \in \mathbb{Z}} 2^{iq}\left(\mu(E_{2^i}) - \mu(E_{2^{i+1}})\right) \lesssim \|h_{\mu,p}\| \, \|f\|^q_{W^{1,p}(\mathbb{G}^n)},$$

which combined with (4.10) further gives us that

$$\int_{\mathbb{R}^n} |f|^q \, d\mu \lesssim \|h_{\mu,p}\| \, \|f\|^q_{W^{1,p}(\mathbb{G}^n)}.$$

Thus (ii) holds with

$$C \lesssim \|h_{\mu,p}\|^{\frac{1}{q}}.$$

(ii) $\Rightarrow$ (i) According to Theorem 3.2.6, we may replace Cap_p in the definition of $h_{\mu,p}$ by $\mathrm{Cap}_{0,p}$. Observe that if $\mu(\mathbb{R}^n) \leq t$ then we take it for granted that $h_{\mu;p}(t) = \infty$.
Thus

$$\|h_{\mu,p}\| = \left(\int_0^{\mu(\mathbb{R}^n)} \frac{ds^{\frac{p}{p-q}}}{(h_{\mu,p}(s))^{\frac{q}{p-q}}} \right)^{\frac{p-q}{p}}.$$

For $\mu(\mathbb{R}^n) < \infty$, there exists a unique $J_0 \in \mathbb{Z}$ such that

$$2^{J_0} < \mu(\mathbb{R}^n) \leq 2^{J_0+1}.$$

For $\mu(\mathbb{R}^n) = \infty$, let $J_0 = \infty$. Since $h_{\mu,p}$ is an increasing function, it follows that

(4.14)
$$\|h_{\mu,p}\| = \left(\sum_{j=-\infty}^{J_0} \int_{2^j}^{2^{j+1}} \frac{ds^{\frac{p}{p-q}}}{(h_{\mu,p}(s))^{\frac{q}{p-q}}} \right)^{\frac{p-q}{p}}$$

$$\leq (2^{\frac{p}{p-q}} - 1)^{\frac{p-q}{p}} \left(\sum_{j=-\infty}^{J_0} \frac{2^{j\frac{p}{p-q}}}{(h_{\mu,p}(2^j))^{\frac{q}{p-q}}} \right)^{\frac{p-q}{p}}.$$

For each $j \in \mathbb{Z}$ such that $j < J_0 + 1$, by the definition of $h_{\mu,p}$ and Theorem 3.2.6, for any $\epsilon \in (0, \infty)$, there exists a compact set $K_j \subseteq \mathbb{R}^n$ such that

$$\mu(K_j) \geq 2^j$$

and

$$\mathrm{Cap}_{0,p}(K_j; \mathbb{G}^n) - \epsilon h_{\mu,p}(2^j) \leq h_{\mu,p}(2^j) \leq \mathrm{Cap}_{0,p}(K_j; \mathbb{G}^n).$$

Further, according to the definition of $\mathrm{Cap}_{0,p}(K_j; \mathbb{G}^n)$, there exists a function

$$f_j \in \mathcal{A}(K_j)$$

such that

(4.15)
$$\|f_j\|^p_{W^{1,p}(\mathbb{G}^n)} - 2^{-j}\epsilon \leq \mathrm{Cap}_{0,p}(K_j; \mathbb{G}^n) \leq \|f_j\|^p_{W^{1,p}(\mathbb{G}^n)}.$$

We define the function

$$F_{i,m} := \max_{i \leq j \leq m} \gamma_j f_j \quad \text{with} \quad \gamma_j := \left(\frac{2^j}{h_{\mu,p}(2^j)} \right)^{\frac{1}{p-q}}.$$

Notice that

$$f_j \in C^0(\mathbb{R}^n) \Rightarrow F_{i,m} \in C^0(\mathbb{R}^n).$$

So, applying

$$f_j \in W^{1,p}(\mathbb{G}^n)$$

and Lemma 3.1.2 yields that $\nabla F_{i,m}$ exists a. e. on $\mathbb{R}^n$ and that

$$|\nabla F_{i,m}| \leq \max_{i \leq j \leq m} \gamma_j |\nabla f_j| \quad a.\,e.\text{ on } \mathbb{R}^n.$$

It is easy to verify

$$\|F_{i,m}\|_{W^{1,p}(\mathbb{G}^n)} \leq \sum_{j=i}^{m} \gamma_j \|f_j\|_{W^{1,p}(\mathbb{G}^n)} < \infty.$$

Hence

$$F_{i,m} \in W^{1,p}(\mathbb{G}^n).$$

Accordingly, (ii) yields

(4.16)
$$\int_{\mathbb{R}^n} |F_{i,m}|^q \, d\mu \leq C^q \|F_{i,m}\|_{W^{1,p}(\mathbb{G}^n)}^q.$$

For the left side of (4.16), using the nonincreasing rearrangement of $F_{i,m}$ implies

$$\int_{\mathbb{R}^n} |F_{i,m}|^q \, d\mu = \int_0^\infty \left(\inf\{s > 0 : \mu(\{x \in \mathbb{R}^n : |F_{i,m}(x)| > s\}) \leq t\} \right)^q dt$$

(4.17)
$$= \sum_{j \in \mathbb{Z}} \int_{2^{j-1}}^{2^j} \left(\inf\{s > 0 : \mu(\{x \in \mathbb{R}^n : |F_{i,m}(x)| > s\}) \leq t\} \right)^q dt$$

$$\geq \sum_{j=i}^{m} 2^{j-1} \left(\inf\{s > 0 : \mu(\{x \in \mathbb{R}^n : |F_{i,m}(x)| > s\}) \leq 2^j\} \right)^q.$$

For every $x \in K_j$ with $i \leq j \leq m$, we have

$$F_{i,m}(x) \geq \gamma_j f_j(x) \geq \gamma_j,$$

thereby getting via (4.3) that for any small number $\eta > 0$,

$$\mu(\{x \in \mathbb{R}^n : |F_{i,m}(x)| > \gamma_j - \eta\}) \geq \mu(K_j) \geq 2^j,$$

so that

$$\inf\left\{s > 0 : \mu(\{x \in \mathbb{R}^n : |F_{i,m}(x)| > s\}) \leq 2^j\right\} \geq \gamma_j - \eta.$$

Letting $\eta \to 0$ gives

$$\inf\left\{s > 0 : \mu(\{x \in \mathbb{R}^n : |F_{i,m}(x)| > s\}) \leq 2^j\right\} \geq \gamma_j \quad \forall \;\; i \leq j \leq m.$$

This, plus (4.17), implies

(4.18)
$$\int_{\mathbb{R}^n} |F_{i,m}|^q \, d\mu \geq \sum_{j=i}^{m} 2^{j-1} \gamma_j^q = 2^{-1} \sum_{j=i}^{m} \frac{2^{j \frac{p}{p-q}}}{\left(h_{\mu,p}(2^j)\right)^{\frac{q}{p-q}}}.$$

Now we compute the right side of (4.16). Clearly,

$$
\|F_{i,m}\|^q_{W^{1,p}(\mathbb{G}^n)} \leq \left(\|F_{i,m}\|^p_{L^p(\mathbb{G}^n)} + \|\nabla F_{i,m}\|^p_{L^p(\mathbb{G}^n)}\right)^{\frac{q}{p}}
$$

$$
\leq \left(\sum_{j=i}^{m} \gamma_j^p \|f_j\|^p_{L^p(\mathbb{G}^n)} + \sum_{j=i}^{m} \gamma_j^p \|\nabla f_j\|^p_{L^p(\mathbb{G}^n)}\right)^{\frac{q}{p}}
$$

$$
= \left(\sum_{j=i}^{m} \gamma_j^p \|f_j\|^p_{W^{1,p}(\mathbb{G}^n)}\right)^{\frac{q}{p}}.
$$

By using (4.3) and (4.15), together with the definition of γ_j, we see

$$
\sum_{j=i}^{m} \gamma_j^p \|f_j\|^p_{W^{1,p}(\mathbb{G}^n)} \leq \sum_{j=i}^{m} \gamma_j^p h_{\mu,p}(2^j) \leq (1+2\epsilon) \sum_{j=i}^{m} \frac{2^{j\frac{p}{p-q}}}{\left(h_{\mu,p}(2^j)\right)^{\frac{q}{p-q}}},
$$

whence

(4.19) $$
\|F_{i,m}\|^q_{W^{1,p}(\mathbb{G}^n)} \leq \left((1+2\epsilon) \sum_{j=i}^{m} \frac{2^{j\frac{p}{p-q}}}{\left(h_{\mu,p}(2^j)\right)^{\frac{q}{p-q}}}\right)^{\frac{q}{p}}.
$$

Inserting the estimates of (4.18) and (4.19) into (4.16), we obtain

$$
2^{-1} \sum_{j=i}^{m} \frac{2^{j\frac{p}{p-q}}}{\left(h_{\mu,p}(2^j)\right)^{\frac{q}{p-q}}} \leq C^q \left((1+2\epsilon) \sum_{j=i}^{m} \frac{2^{j\frac{p}{p-q}}}{\left(h_{\mu,p}(2^j)\right)^{\frac{q}{p-q}}}\right)^{\frac{q}{p}}.
$$

Since $q < p$, it follows that

$$
\left(\sum_{j=i}^{m} \frac{2^{j\frac{p}{p-q}}}{\left(h_{\mu,p}(2^j)\right)^{\frac{q}{p-q}}}\right)^{\frac{p-q}{p}} \leq 2C^q(1+2\epsilon)^{\frac{q}{p}}.
$$

Taking supremum over all $i, m \in \mathbb{Z}$ such that

$$
-\infty < i \leq m < J_0 + 1,
$$

we use (4.14) to deduce

$$
\|h_{\mu,p}\| \leq C_{p,q} \left(\sum_{j=-\infty}^{J_0} \frac{2^{j\frac{p}{p-q}}}{\left(h_{\mu,p}(2^j)\right)^{\frac{q}{p-q}}}\right)^{\frac{p-q}{p}} \leq 2C_{p,q}C^q(1+2\epsilon)^{\frac{q}{p}}.
$$

Letting $\epsilon \to 0$, we obtain (i), thereby completing the proof of the theorem. $\qquad\square$

Chapter 5
Gaussian 1-Capacity to Gaussian ∞-Capacity

The development thus far in the previous chapters has not excluded the case $p = 1$, a situation which almost always requires special treatment involving both analysis and geometry. Our objective in this chapter is to reformulate the Gaussian 1-capacity, characterize the Gaussian Poincaré 1-inequality and Ehrhard's inequality as well as the Gaussian isoperimetry, and handle the Gaussian ∞-capacity as the dual form of the Gaussian 1-capacity.

5.1 Gaussian Co-area Formula and 1-Capacity

The general co-area formula for the Gaussian space (see [34, p. 272] for a special form) reads as follows.

Lemma 5.1.1. For

$$(f, g) \in W^{1,1}(\mathbb{G}^n) \times L^\infty(\mathbb{G}^n)$$

one has

(5.1)
$$\int_{\mathbb{R}^n} |\nabla f| g \, dV_\gamma = \int_0^\infty \left(\int_{\{x \in \mathbb{R}^n : |f(x)| = s\}} g(x)\gamma(x) \, dP(x) \right) ds,$$

where on the left side the convention $0 \cdot \infty = 0$ is utilized and on the right side the symbol dP stands for the Hausdorff measure of dimension $n - 1$ on the surface

$$\{x \in \mathbb{R}^n : |f(x)| = s\}.$$

Proof. Following the argument for [11, Lemma 4.1] we choose

$$\psi_R(x) = 1_{B(0,R)}(x) + \left(\frac{R + n - |x|}{n} \right)^2 1_{B(0,R+n)\backslash B(0,R)}(x) \quad \forall \ (x, R) \in \mathbb{R}^n \times (0, \infty),$$

whence getting

$$\sup_{(x,R) \in \mathbb{R}^n \times (0,\infty)} |\nabla \psi_R(x)| \le 1.$$

Clearly, if

$$f \in W^{1,1}(\mathbb{G}^n) \ \& \ f_j = \psi_j f \ \forall \ j \in \mathbb{N},$$

then f_j converges to f in $W^{1,1}(\mathbb{G}^n)$ as $j \to \infty$. If g is compactly supported, then for j large enough one has

$$|\nabla f_j| g = |\nabla f| g,$$

© Springer Nature Switzerland AG 2018
L. Liu et al., *Gaussian Capacity Analysis*, Lecture Notes in Mathematics 2225,
https://doi.org/10.1007/978-3-319-95040-2_5

and hence the formula (5.1) is valid when g has compact support in $\mathbb{G}^n$. Of course, the general case of (5.1) follows by the dominated convergence theorem due to the fact that any function $g \in L^\infty(\mathbb{G}^n)$ can be approximated by the function $g1_{B(0,j)}$.   □

For every Borel set $A \subseteq \mathbb{R}^n$, define the Gaussian Minkowski content of its boundary ∂A as

$$P_\gamma(A) := \liminf_{r \to 0} r^{-1}(V_\gamma(A_r) - V_\gamma(A)),$$

where

$$A_r := \{x \in \mathbb{R}^n : \operatorname{dist}(x, A) \le r\}.$$

If the Borel set $A \subseteq \mathbb{R}^n$ has smooth boundary ∂A, then

$$P_\gamma(A) = \int_{\partial A} \gamma(x)\, dP(x).$$

According to [34, (8.25)], for any Borel set $A \subseteq \mathbb{R}^n$ with smooth boundary ∂A one has

$$\frac{P_\gamma(A)}{V_\gamma(A)V_\gamma(\mathbb{R}^n \setminus A)} \ge 2\sqrt{\frac{2}{\pi}}.$$

This inequality may be interpreted as Cheeger's isoperimetric inequality on $\mathbb{G}^n$; see also [34, 13].

Lemma 5.1.1 is now used to prove the following geometric characterization of the Gaussian 1-capacity.

Theorem 5.1.2. If K is a compact subset of $\mathbb{R}^n$, then

$$\operatorname{Cap}_1(K; \mathbb{G}^n) = \operatorname{Cap}_{1,*}(K; \mathbb{G}^n)$$

whose right side is determined by

$$\inf\left\{P_\gamma(O) + V_\gamma(O) : \text{open } O \supseteq K \text{ with compact } \overline{O} \text{ and smooth } \partial O\right\}.$$

Proof. Let K be a compact subset of $\mathbb{R}^n$. From (3.22) and Remark 3.2.2 it follows that

$$\operatorname{Cap}_1(K; \mathbb{G}^n) = \inf\left\{\|f\|_{W^{1,1}(\mathbb{G}^n)}^p : 0 \le f \in \mathcal{A}(K)\right\}.$$

Using the Gaussian co-area formula in Lemma 5.1.1, we know that if

$$0 \le f \in \mathcal{A}(K)$$

then

$$\|f\|_{W^{1,1}(\mathbb{G}^n)} = \int_{\mathbb{R}^n} |\nabla f|\, dV_\gamma + \int_{\mathbb{R}^n} |f|\, dV_\gamma$$

$$\ge \int_0^1 \left(P_\gamma(\{x \in \mathbb{R}^n : f(x) = s\}) + V_\gamma(\{x \in \mathbb{R}^n : f(x) > s\})\right) ds$$

$$\geq \mathrm{Cap}_{1,*}(K; \mathbb{G}^n)$$

and hence

$$\mathrm{Cap}_1(K; \mathbb{G}^n) \geq \mathrm{Cap}_{1,*}(K; \mathbb{G}^n).$$

In order to verify the converse, suppose that O is an open set used in the definition of $\mathrm{Cap}_{1,*}(K; \mathbb{G}^n)$,

$$\mathrm{dist}(x, \mathbb{R}^n \setminus O)$$

is the Euclidean distance from x to $\mathbb{R}^n \setminus O$, and

$$O_t := \left\{ x \in \mathbb{R}^n : \mathrm{dist}(x, \mathbb{R}^n \setminus O) > t \right\} \quad \forall \quad t \in (0, \infty).$$

Let $\epsilon \in (0, 1)$. Choosing an increasing C^∞-function $\phi_\epsilon : [0, \infty) \to [0, \infty)$ with

$$\phi_\epsilon(s) = \begin{cases} 0 & \forall \quad s \in [0, \epsilon); \\ 1 & \forall \quad s \in [2\epsilon, \infty), \end{cases}$$

defining

$$f_O(x) := \phi_\epsilon\big(\mathrm{dist}(x, \mathbb{R}^n \setminus O)\big) \quad \forall \quad x \in \mathbb{R}^n,$$

and using the Gaussian co-area formula in Lemma 5.1.1, we see

$$\int_{\mathbb{R}^n} |\nabla f_O| \, dV_\gamma = \int_0^{2\epsilon} P_\gamma(O_t) \phi_\epsilon'(t) \, dt$$

$$= \int_0^{2\epsilon} \big(P_\gamma(O_t) - P_\gamma(O)\big) \phi_\epsilon'(t) \, dt + P_\gamma(O)$$

$$\to P_\gamma(O) \quad \text{as} \quad \epsilon \to 0.$$

Notice that

$$\int_{\mathbb{R}^n} f_O \, dV_\gamma = \int_O f_O \, dV_\gamma \to V_\gamma(O) \quad \text{as} \quad \epsilon \to 0.$$

So we obtain

$$\mathrm{Cap}_1(K; \mathbb{G}^n) \leq P_\gamma(O) + V_\gamma(O),$$

whence

$$\mathrm{Cap}_1(K; \mathbb{G}^n) \leq \mathrm{Cap}_{1,*}(K; \mathbb{G}^n).$$

This completes the proof of the theorem.   □

5.2 Gaussian Poincaré 1-Inequality

Next, we utilize Lemma 5.1.1 again to discover an optimal relationship between Cheeger's isoperimetric inequality on $\mathbb{G}^n$ and the Gaussian Poincaré 1-inequality, thereby reshowing the geometric case $p = 1$ of $(\star)$.

Theorem 5.2.1. The following three statements are true and equivalent:

(i) For any open set $O \subseteq \mathbb{R}^n$ with smooth boundary one has

$$(5.2) \qquad \frac{P_\gamma(O)}{2\sqrt{\frac{2}{\pi}}} \geq V_\gamma(O)V_\gamma(\mathbb{R}^n \setminus O).$$

(ii) For any function

$$f \in C^1(\mathbb{R}^n)$$

one has

$$(5.3) \qquad \frac{\int_{\mathbb{R}^n} |\nabla f| \, dV_\gamma}{2\sqrt{\frac{2}{\pi}}} \geq \int_{-\infty}^{\infty} \frac{V_\gamma(\{x \in \mathbb{R}^n : f(x) > s\})}{\left(V_\gamma(\{x \in \mathbb{R}^n : f(x) \leq s\})\right)^{-1}} \, ds.$$

(iii) For any function $f \in C^1(\mathbb{R}^n)$ with

$$\int_{\mathbb{R}^n} f \, dV_\gamma = 0$$

one has

$$(5.4) \qquad \frac{\int_{\mathbb{R}^n} |\nabla f| \, dV_\gamma}{\sqrt{\frac{2}{\pi}}} \geq \int_{\mathbb{R}^n} |f| \, dV_\gamma.$$

Proof. The argument comprises two parts.

Part 1 - proving (5.2).

Following the proof of [34, Proposition 8.5], for two smooth functions f, g on $\mathbb{R}^n$, a Borel set $O \subseteq \mathbb{R}^n$ with smooth boundary ∂O and a nonnegative number t we compute

$$\int_{\mathbb{R}^n} g(x)(P_t f - f)(x) \, dV_\gamma(x) = -\int_0^t \left(\int_{\mathbb{R}^n} \nabla P_s g(x) \cdot \nabla f(x) \, dV_\gamma(x) \right) ds,$$

where

$$\begin{cases} P_t f(x) = \int_{\mathbb{R}^n} f(e^{-t}x + \sqrt{1 - e^{-2t}}y) \, dV_\gamma(y); \\ \nabla P_s g(x) = \frac{e^{-s}}{\sqrt{1 - e^{-2s}}} \int_{\mathbb{R}^n} yg(e^{-s}x + \sqrt{1 - e^{-2s}}y) \, dV_\gamma(y). \end{cases}$$

Accordingly, we find

$$-\int_{\mathbb{R}^n} g(x)(P_t f - f)(x) \, dV_\gamma(x)$$

$$= \int_0^t \frac{e^{-s}}{\sqrt{1 - e^{-2s}}} \left(\iint_{\mathbb{R}^n \times \mathbb{R}^n} y \cdot \nabla f(x)g(e^{-s}x + \sqrt{1 - e^{-2s}}y) \, dV_\gamma(y) \, dV_\gamma(x) \right) ds,$$

which is still valid for the limit

$$g \to 1_{\mathbb{R}^n \backslash O}.$$

Note that

$$\iint_{\mathbb{R}^n \times \mathbb{R}^n} y \cdot \nabla f(x) \, dV_\gamma(y) \, dV_\gamma(x) = 0.$$

So

$$-\iint_{\mathbb{R}^n \times \mathbb{R}^n} y \cdot \nabla f(x) 1_{\mathbb{R}^n \backslash O} \left(e^{-s} x + \sqrt{1 - e^{-2s}} y \right) dV_\gamma(y) \, dV_\gamma(x)$$

$$\leq 2^{-1} \iint_{\mathbb{R}^n \times \mathbb{R}^n} \left| y \cdot \nabla f(x) \right| dV_\gamma(y) \, dV_\gamma(x)$$

$$= \frac{1}{\sqrt{2\pi}} \int_{\mathbb{R}^n} |\nabla f| \, dV_\gamma$$

$$\to \frac{P_\gamma(O)}{\sqrt{2\pi}} \quad \text{as } f \to 1_O.$$

Consequently, we achieve

$$\int_O P_t(1_{\mathbb{R}^n \backslash O}) \, dV_\gamma \leq \left(\frac{P_\gamma(O)}{\sqrt{2\pi}} \right) \arccos(e^{-t}) \quad \forall \ t \in [0, \infty),$$

thereby reaching (5.2) via letting $t \to \infty$.

  Part 2 - checking (i)$\Longrightarrow$(ii)$\Longrightarrow$(iii)$\Longrightarrow$(i).

  First of all, we prove that (i) implies (ii). Let f be a smooth function. Then for any $s \in \mathbb{R}$, the set

$$\{ x \in \mathbb{R}^n : f(x) > s \}$$

has smooth boundary. If (i) is valid, then an application of P_γ and (5.1) derives

$$\int_{\mathbb{R}^n} |\nabla f| \, dV_\gamma = \int_{-\infty}^{\infty} \left(\int_{\{x \in \mathbb{R}^n : f(x) = s\}} \gamma(x) \, dP(x) \right) ds$$

$$= \int_{-\infty}^{\infty} P_\gamma(\{ x \in \mathbb{R}^n : f(x) > s \}) \, ds$$

$$\geq 2\sqrt{\frac{2}{\pi}} \int_{-\infty}^{\infty} V_\gamma(\{ x \in \mathbb{R}^n : f(x) > s \}) V_\gamma(\{ x \in \mathbb{R}^n : f(x) \leq s \}) \, ds,$$

whence reaching (ii).

  Next, suppose that (ii) holds. In order to validate (iii), let f be a smooth function with

$$\int_{\mathbb{R}^n} f \, dV_\gamma = 0.$$

Then

$$\int_0^\infty V_\gamma(\{ x \in \mathbb{R}^n : f(x) > s \}) \, ds = \int_0^\infty V_\gamma(\{ x \in \mathbb{R}^n : f(x) < -s \}) \, ds$$

and hence

$$(5.5) \qquad \int_{\mathbb{R}^n} |f| \, dV_\gamma = 2 \int_0^\infty V_\gamma(\{x \in \mathbb{R}^n : f(x) > s\}) \, ds.$$

Now (5.5) implies

$$
\begin{aligned}
\int_{-\infty}^0 & V_\gamma(\{x \in \mathbb{R}^n : f(x) < s\}) \, ds \\
(5.6) \qquad &= \int_0^\infty V_\gamma(\{x \in \mathbb{R}^n : f(x) < -s\}) \, ds \\
&= 2^{-1} \int_{\mathbb{R}^n} |f| \, dV_\gamma.
\end{aligned}
$$

For any $s \in \mathbb{R}$, write

$$
\begin{cases}
\varphi(s) := V_\gamma(\{x \in \mathbb{R}^n : f(x) > s\}); \\
\psi(s) := V_\gamma(\{x \in \mathbb{R}^n : f(x) < s\}).
\end{cases}
$$

Then φ is nonnegative decreasing and ψ is nonnegative increasing on $\mathbb{R}$. Moreover, since

$$V_\gamma(\mathbb{R}^n) = 1,$$

we see

$$\varphi + \psi = 1.$$

Thus, by (5.6), we achieve

$$
\begin{aligned}
\int_{-\infty}^\infty & V_\gamma(\{x \in \mathbb{R}^n : f(x) > s\}) V_\gamma(\{x \in \mathbb{R}^n : f(x) \le s\}) \, ds \\
&= \int_0^\infty \varphi(s)\psi(s) \, ds + \int_{-\infty}^0 \varphi(s)\psi(s) \, ds \\
&\ge \psi(0) \int_0^\infty \varphi(s) \, ds + \varphi(0) \int_{-\infty}^0 \psi(s) \, ds \\
&= \psi(0) \int_0^\infty V_\gamma(\{x \in \mathbb{R}^n : f(x) > s\}) \, ds + (1 - \psi(0)) \int_{-\infty}^0 V_\gamma(\{x \in \mathbb{R}^n : f(x) < s\}) \, ds \\
&= 2^{-1} \int_{\mathbb{R}^n} |f| \, dV_\gamma.
\end{aligned}
$$

This, along with (5.3), yields

$$\int_{\mathbb{R}^n} |\nabla f| \, dV_\gamma \ge \sqrt{\frac{2}{\pi}} \int_{\mathbb{R}^n} |f| \, dV_\gamma.$$

Thus, (iii) follows.

Finally, we prove that (iii) implies (i). To this end, let $O \subseteq \mathbb{R}^n$ be an open set with smooth boundary and $O^c = \mathbb{R}^n \setminus O$ be the complement of O. Define the function f_O as in the proof of Theorem 5.1.2, and consider

$$f_\dagger = f_O - \int_{\mathbb{R}^n} f_O \, dV_\gamma.$$

By the definition of f_O, we have

$$f_O(x) = \begin{cases} 1 & \forall \ x \in O_{2\epsilon}; \\ 0 & \forall \ x \notin O_\epsilon \end{cases}$$

and

$$f_\dagger(x) = \begin{cases} 1 - V_\gamma(O_{2\epsilon}) - \int_{O_\epsilon \setminus O_{2\epsilon}} f_O \, dV_\gamma & \forall \ x \in O_{2\epsilon}; \\ - \int_{\mathbb{R}^n} f_O \, dV_\gamma & \forall \ x \notin O_\epsilon. \end{cases}$$

Hence

$$\int_{\mathbb{R}^n} |f_\dagger| \, dV_\gamma = I + II + III,$$

where

$$\begin{cases} I := \int_{O_{2\epsilon}} \left| 1 - V_\gamma(O_{2\epsilon}) - \int_{O_\epsilon \setminus O_{2\epsilon}} f_O \, dV_\gamma \right| dV_\gamma; \\ II := \int_{O_\epsilon \setminus O_{2\epsilon}} |f| \, dV_\gamma; \\ III := V_\gamma(O_\epsilon^c) \int_{\mathbb{R}^n} f_O \, dV_\gamma. \end{cases}$$

Notice that

$$I \geq V_\gamma(O_{2\epsilon}) V_\gamma(O_{2\epsilon}^c) - V_\gamma(O_{2\epsilon}) \left| \int_{O_\epsilon \setminus O_{2\epsilon}} f_O \, dV_\gamma \right| \to V_\gamma(O) V_\gamma(O^c) \quad \text{as} \quad \epsilon \to 0.$$

Also,

$$II \to 0 \quad \text{as} \quad \epsilon \to 0.$$

Noticing that

$$\int_{\mathbb{R}^n} f_O \, dV_\gamma = \int_{O_\epsilon} f_O \, dV_\gamma \to V_\gamma(O) \quad \text{as} \quad \epsilon \to 0,$$

we have

$$III \to V_\gamma(O^c) V_\gamma(O) \quad \text{as} \quad \epsilon \to 0.$$

Thus, upon letting $\epsilon \to 0$ we have

$$\int_{\mathbb{R}^n} |f_\dagger| \, dV_\gamma \geq 2 V_\gamma(O) V_\gamma(O^c).$$

Again, a further application of (5.1) implies that

$$\int_{\mathbb{R}^n} |\nabla f_\dagger| \, dV_\gamma = \int_{\mathbb{R}^n} |\nabla f_O| \, dV_\gamma$$

$$= \int_0^{2\epsilon} P_\gamma(O_t)\phi'(t)\,dt$$

$$\to P_\gamma(O) \quad \text{as} \quad \epsilon \to 0.$$

Via combining the last two formulas and applying (5.4), we arrive at (5.2), thereby finding the truth of (i). $\qquad\qquad\square$

5.3 Ehrhard's Inequality and Gaussian Isoperimetry

This section is used to show the Ehrhard inequality and its straightforward consequence - the Gaussian isoperimetric inequality.

Theorem 5.3.1. For

$$(\lambda, t) \in (0,1) \times \mathbb{R} \ \& \ A, B \subseteq \mathbb{R}^n$$

let

$$\begin{cases} \lambda A + (1-\lambda)B = \{\lambda x + (1-\lambda)y : (x,y) \in A \times B\}; \\ \Phi(t) = (2\pi)^{-\frac{1}{2}} \int_{-\infty}^t \exp\left(-\frac{s^2}{2}\right) ds; \\ \Psi(t) = \Phi' \circ \Phi^{-1}(t). \end{cases}$$

Then

(i) Ehrhard's inequality

$$\Phi^{-1}\left(V_\gamma\big(\lambda A + (1-\lambda)B\big)\right) \geq \lambda \Phi^{-1}\left(V_\gamma(A)\right) + (1-\lambda)\Phi^{-1}\left(V_\gamma(B)\right)$$

holds for any $\lambda \in (0,1)$ and all Borel sets $A, B \subseteq \mathbb{R}^n$.

(ii) Gaussian isoperimetry

$$\Psi\big(V_\gamma(B)\big) \leq P_\gamma(B)$$

is true for any Borel set $B \subseteq \mathbb{R}^n$.

Proof. (i) The argument is taken from the heat-equation-based proof of [9, Theorem 1.1] whose special cases can be found in [17] for convex bodies $A, B \subseteq \mathbb{R}^n$ and in [32] for any convex $A \subseteq \mathbb{R}^n$ and any Borel set $B \subseteq \mathbb{R}^n$.

First of all, let

$$\Delta u(x,t) = \operatorname{div}\big(\nabla u(x,t)\big) = \left(\frac{\partial^2}{\partial x_1^2} + \cdots + \frac{\partial^2}{\partial x_n^2}\right) u(x,t) \quad \text{for} \quad x = (x_1, \ldots, x_n).$$

If $u(x,t)$ is a positive solution of the following half-heat equation

$$\frac{\partial}{\partial t} u(x,t) = 2^{-1} \Delta u(x,t),$$

then the inverse Gaussian transformation

$$U(x, t) = \Phi^{-1}\big(u(x, t)\big)$$

enjoys

(5.7)
$$\begin{cases} \nabla u(x, t) = \Phi'\big(U(x, t)\big)\nabla U(x, t); \\ \Delta u(x, t) = \Phi'\big(U(x, t)\big)\Big(\Delta U(x, t) - U(x, t)|\nabla U(x, t)|^2\Big); \\ \frac{\partial}{\partial t}U(x, t) = 2^{-1}\Big(\Delta U(x, t) - U(x, t)|\nabla U(x, t)|^2\Big). \end{cases}$$

Note that if U satisfies the half-heat equation, then so does $-U$, and if

$$U(x, 0) = ax + b$$

for two real constants a and b, then

$$U(x, t) = \frac{ax}{\sqrt{1 + a^2 t}} + \frac{b}{\sqrt{1 + a^2 t}}$$

solves the half-heat equation.

Next, without loss of generality we may assume that A and B are nonempty compact subsets of $\mathbb{R}^n$, fix $\epsilon \in (0, 1)$ and choose an infinitely differentiable function

$$F \in C^\infty(\mathbb{R}^n)$$

such that

$$\begin{cases} 0 \leq F \leq 1; \\ F(x) = 1 \quad \forall \quad x \in A; \\ F(x) = 0 \quad \forall \quad x \notin A_\epsilon = A + \{z \in \mathbb{R}^n : |z| \leq \epsilon\}. \end{cases}$$

For $0 < \delta < \epsilon$ set

$$\alpha = \delta + 1 - \epsilon \quad \& \quad f = \delta + (1 - \epsilon)F.$$

Then

$$\begin{cases} f \in C^\infty(\mathbb{R}^n); \\ \delta \leq f \leq \alpha; \\ f(x) = \alpha \quad \forall \quad x \in A; \\ f(x) = \delta \quad \forall \quad x \notin A_\epsilon. \end{cases}$$

Analogously, upon replacing $F; A; A_\epsilon$ by

$$G; B; B_\epsilon = B + \{z \in \mathbb{R}^n : |z| \leq \epsilon\}$$

respectively we achieve a function

$$g = \delta + (1 - \epsilon)G \in C^\infty(\mathbb{R}^n)$$

such that

$$\begin{cases} \delta \le g \le \alpha; \\ g(x) = \alpha \quad \forall \quad x \in B; \\ g(x) = \delta \quad \forall \quad x \notin B_\epsilon. \end{cases}$$

Via letting

$$\kappa(\alpha, \delta) := \max \left\{ \Phi\big(\lambda\Phi^{-1}(\alpha) + (1-\lambda)\Phi^{-1}(\delta)\big), \Phi\big(\lambda\Phi^{-1}(\delta) + (1-\lambda)\Phi^{-1}(\alpha)\big) \right\}$$

we obtain

$$\lim_{\delta \to 0} \kappa(\alpha, \delta) = 0.$$

Now, it is natural to choose a function

$$h \in C^\infty(\mathbb{R}^n)$$

such that

$$\begin{cases} \kappa(\alpha, \delta) \le h \le \alpha; \\ h(x) = \alpha \quad \forall \quad x \in \lambda A_\epsilon + (1-\lambda)B_\epsilon; \\ h(x) = \kappa(\alpha, \delta) \quad \forall \quad x \notin \big(\lambda A_\epsilon + (1-\lambda)B_\epsilon\big)_\epsilon. \end{cases}$$

As a result, we find

$$\Phi^{-1}\big(h(\lambda x + (1-\lambda)y)\big) \ge \lambda\Phi^{-1}\big(f(x)\big) + (1-\lambda)\Phi^{-1}\big(g(y)\big) \quad \forall \quad (x,y) \in \mathbb{R}^n \times \mathbb{R}^n.$$

If the following inequality

$$\Phi^{-1}\left(\int_{\mathbb{R}^n} h \, dV_\gamma\right) \ge \lambda\Phi^{-1}\left(\int_{\mathbb{R}^n} f \, dV_\gamma\right) + (1-\lambda)\Phi^{-1}\left(\int_{\mathbb{R}^n} g \, dV_\gamma\right)$$

is verified, then the desired Ehrhard inequality follows from letting $\delta \to 0$ and then $\epsilon \to 0$ in the last inequality. To reach the last inequality, it suffices to consider

$$u_\rho(x,t) = \int_{\mathbb{R}^n} \rho(x + \sqrt{t}z) \, dV_\gamma(z) \quad (\rho, x, t) \in \{f, g, h\} \times \mathbb{R}^n \times [0, \infty)$$

and show that

(5.8) $$\Phi^{-1}\big(u_h(\lambda x + (1-\lambda)y), t\big) \ge \lambda\Phi^{-1}\big(u_f(x,t)\big) + (1-\lambda)\Phi^{-1}\big(u_g(y,t)\big)$$

holds for any $(x, y, t) \in \mathbb{R}^n \times \mathbb{R}^n \times [0, \infty)$.

Regarding (5.8), we consider the inverse Gaussian transformation

$$U_\rho(x,t) = \Phi^{-1}\big(u_\rho(x,t)\big)$$

of u_ρ, thereby getting

(5.9) $$\begin{cases} \sup_{(x,t) \in \mathbb{R}^n \times [0,\infty)} |U_\rho(x,t)| < \infty; \\ \sup_{(x,t) \in \mathbb{R}^n \times [0,\infty)} \left| \frac{\partial^{i_1 + \cdots + i_n}}{\partial x^{i_1} \cdots \partial x^{i_n}} U_\rho(x,t) \right| < \infty \quad \forall \quad i_1, \dots, i_n \in \mathbb{N}. \end{cases}$$

We now define

$$\Upsilon(x, y, t) = U_h(\lambda x + (1 - \lambda)y, t) - \lambda U_f(x, t) - (1 - \lambda)U_g(y, t)$$

for any

$$(x, y, t) \in \mathbb{R}^n \times \mathbb{R}^n \times [0, \infty).$$

So, the above desired inequality follows from validating

(5.10) $\qquad \Upsilon(x, y, t) \geq 0 \quad \forall \quad (x, y, t) \in \mathbb{R}^n \times \mathbb{R}^n \times (0, \infty).$

Finally, for simplicity, let

$$\begin{cases} \zeta = (x, t); \\ \eta = (y, t); \\ \theta = (\lambda x + (1 - \lambda)y, t). \end{cases}$$

Then

(5.11) $\qquad \begin{cases} \nabla_x \Upsilon = \lambda((\nabla U_h)(\theta) - (\nabla U_f)(\zeta)); \\ \nabla_y \Upsilon = (1 - \lambda)((\nabla U_h)(\theta) - (\nabla U_f)(\eta)); \\ \Delta_x \Upsilon = \lambda^2 (\Delta U_h)(\theta) - \lambda(\Delta U_f)(\zeta); \\ \Delta_y \Upsilon = (1 - \lambda)^2 (\Delta U_h)(\theta) - (1 - \lambda)(\Delta U_f)(\zeta); \\ \sum_{1 \leq i \leq n} \frac{\partial^2 \Upsilon(x, y, t)}{\partial x_i \partial y_i} = \lambda(1 - \lambda)(\Delta U_h)(\theta). \end{cases}$

Consequently, upon introducing the differential operator

$$\mathscr{D} = 2^{-1} \left(\Delta_x + 2 \sum_{1 \leq i \leq n} \frac{\partial^2}{\partial x_i \partial y_i} + \Delta_y \right)$$

we utilize (5.7) to compute

$$\mathscr{D}\Upsilon(x, y, t) = 2^{-1} \left((\Delta U_h)(\theta) - \lambda(\Delta U_f)(\zeta) - (1 - \lambda)(\Delta U_g)(\eta) \right)$$

$$= \partial_t U_h(\theta) + 2^{-1} U_h(\theta)|(\nabla U_h)(\theta)|^2 - \lambda \partial_t U_f(\zeta) - 2^{-1}\lambda U_f(\zeta)|(\nabla U_f)(\zeta)|^2$$

$$- (1 - \lambda)\partial_t U_g(\eta) - 2^{-1}(1 - \lambda)U_g(\eta)|(\nabla U_g)(\eta)|^2$$

$$= \partial_t \Upsilon(x, y, t) + \Xi(x, y, t),$$

where

$$\begin{cases} \Xi(x, y, t); \\ |(\nabla U_f)(\zeta)|^2; \\ |(\nabla U_g)(\eta)|^2, \end{cases}$$

are defined by

$$\begin{cases} 2^{-1}\Big(U_h(\theta)|(\nabla U_h)(\theta)|^2 - \lambda U_f(\zeta)|(\nabla U_f)(\zeta)|^2 - (1-\lambda)U_g(\eta)|(\nabla U_g)(\eta)|^2\Big); \\ |(\nabla U_h)(\theta)|^2 + \sum_{1 \le i \le n}\Big(\partial_{x_i}U_f(\zeta) + \partial_{x_i}U_h(\theta)\Big)\Big(\partial_{x_i}U_f(\zeta) - \partial_{x_i}U_h(\theta)\Big); \\ |(\nabla U_h)(\theta)|^2 + \sum_{1 \le i \le n}\Big(\partial_{x_i}U_g(\eta) + \partial_{x_i}U_h(\theta)\Big)\Big(\partial_{x_i}U_g(\eta) - \partial_{x_i}U_h(\theta)\Big), \end{cases}$$

respectively. Accordingly, (5.11) produces a continuous function $\beta(x,y,t)$ (which depends on (5.9) and is Lipschitz continuous in (x,y) with a Lipschitz constant being uniformly bounded in $[0,\infty)$) such that

$$\Xi(x,y,t) = 2^{-1}|(\nabla U_h)(\theta)|^2\Upsilon(x,y,t) - \beta(x,y,t) \cdot \nabla_{(x,t)}\Upsilon(x,y,t)$$

and

$$\mathscr{D}\Upsilon(x,y,t) + \beta(x,y,t) \cdot \nabla_{(x,y)}\Upsilon(x,y,t) = \partial_t\Upsilon(x,y,t) + 2^{-1}|(\nabla U_h)(\theta)|^2\Upsilon(x,y,t),$$

where $\nabla_{(x,y)}$ is the gradient in (x,y).

Given $T \in (0,\infty)$. Since the definitions of the above functions f, g, h derive

$$\liminf_{|x|+|y|\to\infty} \inf_{0 \le t \le T} \Upsilon(x,y,t) \ge 0,$$

it follows that if

$$\Upsilon(x_*,y_*,t_*) < 0 \quad \text{at some} \quad (x_*,y_*,t_*) \in \mathbb{R}^n \times \mathbb{R}^n \times [0,T],$$

then there exists an $\epsilon_0 > 0$ such that the function $\epsilon_0 t + \Upsilon(x,y,t)$ has a strictly negative minimum in $\mathbb{R}^n \times \mathbb{R}^n \times [0,T]$ at some point

$$P_0 = (x_0, y_0, t_0) \quad \text{with} \quad t_0 > 0.$$

This in turn implies

$$\begin{cases} \Upsilon(P_0) < 0; \\ \partial_t\Upsilon(P_0) \le -\epsilon_0; \\ \nabla_{(x,y)}\Upsilon(P_0) = 0; \\ \mathscr{D}\Upsilon(P_0) \ge 0, \end{cases}$$

and so

$$\mathscr{D}\Upsilon(P_0) + \beta(P_0) \cdot \nabla_{(x,y)}\Upsilon(P_0) \ge 0 > -\epsilon_0 > \partial_t\Upsilon(P_0) + 2^{-1}|(\nabla U_h)(\theta)|^2\Upsilon(P_0).$$

However, this strict inequality is against the known equality

$$\mathscr{D}\Upsilon(P_0) + \beta(P_0) \cdot \nabla_{(x,y)}\Upsilon(P_0) = \partial_t\Upsilon(P_0) + 2^{-1}|(\nabla U_h)(\theta)|^2\Upsilon(P_0).$$

Accordingly, we must have the desired nonnegativity (5.9).

(ii) With the just-verified (i), we have that if B is a Borel subset of $\mathbb{R}^n$ and

$$B_t = \{x \in \mathbb{R}^n : |x - x_0| < t \text{ for some } x_0 \in B\} = B + tB(0, 1)$$

is its $0 < t$-enlargement, then

$$\Phi^{-1}\big(V_\gamma(B_t)\big) = \Phi^{-1}\Big(V_\gamma\big(\lambda(\lambda^{-1}B) + (1 - \lambda)((1 - \lambda)^{-1}tB(0, 1))\big)\Big)$$

$$\geq \lambda\Phi^{-1}\Big(V_\gamma\big(\lambda^{-1}B\big)\Big) + (1 - \lambda)\Phi^{-1}\Big(V_\gamma\big((1 - \lambda)^{-1}tB(0, 1)\big)\Big)$$

$$\to \Phi^{-1}\big(V_\gamma(B)\big) + t \quad \text{as} \quad (0, 1) \ni \lambda \to 1,$$

whence

$$V_\gamma(B_t) \geq \Phi\Big(t + \Phi^{-1}\big(V_\gamma(B)\big)\Big) \quad \forall \quad t > 0.$$

This amounts to the Gaussian isoperimetric inequality described in (ii). $\qquad\square$

5.4 Gaussian ∞-Capacity

Note that $L^\infty(\mathbb{G}^n)$ is the dual space of $L^1(\mathbb{G}^n)$. So we are required to introduce the following notion of the Gaussian ∞-capacity (as a dual form of $\text{Cap}_1(\cdot; \mathbb{G}^n)$) and then investigate the relations between the Gaussian ∞-capacity and the Gaussian p-capacity.

Definition 5.4.1. For any set $E \subseteq \mathbb{R}^n$ let

$$\mathcal{A}_\infty(E) := \big\{f \in W^{1,\infty}(\mathbb{R}^n) : E \subseteq \{x \in \mathbb{R}^n : f(x) \geq 1\}^\circ\big\}.$$

Define

$$\text{Cap}_\infty(E; \mathbb{G}^n) := \inf \big\{\|f\|_{W^{1,\infty}(\mathbb{G}^n)} : f \in \mathcal{A}_\infty(E)\big\};$$

$$\text{Cap}_\infty^*(E; \mathbb{G}^n) := \inf \big\{\||\nabla f|\|_{L^\infty(\mathbb{G}^n)} + \|f\|_{L^1(\mathbb{G}^n)} : f \in \mathcal{A}_\infty(E)\big\};$$

and

$$\text{Cap}_\infty^{**}(E; \mathbb{G}^n) := \inf \Big\{\||\nabla f|\|_{L^\infty(\mathbb{G}^n)} + \Big|\int_{\mathbb{R}^n} f \, dV_\gamma\Big| : f \in \mathcal{A}_\infty(E)\Big\}.$$

Proposition 5.4.2. For any set $E \subseteq \mathbb{R}^n$ one has

(5.12) $$\lim_{p \to \infty} \big(\text{Cap}_p(E; \mathbb{G}^n)\big)^{\frac{1}{p}} \leq \text{Cap}_\infty(E; \mathbb{G}^n) \leq 2 \lim_{p \to \infty} \big(\text{Cap}_p(E; \mathbb{G}^n)\big)^{\frac{1}{p}}$$

and

(5.13) $$\text{Cap}_\infty(E; \mathbb{G}^n) \simeq \text{Cap}_\infty^*(E; \mathbb{G}^n) \simeq \text{Cap}_\infty^{**}(E; \mathbb{G}^n).$$

Proof. Given any set $E \subseteq \mathbb{R}^n$, by Proposition 3.1.4(iv), we see that

$$p \mapsto 2^{-\frac{1}{p}} \left(\mathrm{Cap}_p(E; \mathbb{G}^n) \right)^{\frac{1}{p}}$$

is a nondecreasing function, which ensures

$$\lim_{p \to \infty} 2^{-\frac{1}{p}} \left(\mathrm{Cap}_p(E; \mathbb{G}^n) \right)^{\frac{1}{p}}$$

exists, so does

$$\lim_{p \to \infty} \left(\mathrm{Cap}_p(E; \mathbb{G}^n) \right)^{\frac{1}{p}}.$$

Also, since the function $1 \in \mathcal{A}_\infty(E)$, we have

$$\max \left\{ \mathrm{Cap}_\infty(E; \mathbb{G}^n), \ \mathrm{Cap}_\infty^*(E; \mathbb{G}^n), \ \mathrm{Cap}_\infty^{**}(E; \mathbb{G}^n) \right\} \le 1.$$

Assuming first that (5.12) holds, we prove (5.13). It is easy to see that

$$\mathrm{Cap}_\infty(E; \mathbb{G}^n) \ge \mathrm{Cap}_\infty^*(E; \mathbb{G}^n) \ge \mathrm{Cap}_\infty^{**}(E; \mathbb{G}^n).$$

By (5.12), the Poincaré inequality (1.11) and

$$\mathcal{A}_\infty(E) \subseteq \mathcal{A}_p(E) \quad \forall \quad p \in [1, \infty),$$

we obtain

$$
\begin{aligned}
\mathrm{Cap}_\infty(E; \mathbb{G}^n) &\le 2 \lim_{p \to \infty} \left(\mathrm{Cap}_p(E; \mathbb{G}^n) \right)^{\frac{1}{p}} \\
&= 2 \lim_{p \to \infty} \inf_{f \in \mathcal{A}_p(E)} \|f\|_{W^{1,p}(\mathbb{G}^n)} \\
&\simeq \lim_{p \to \infty} \inf_{f \in \mathcal{A}_p(E)} \left\{ \|| \nabla f \|\|_{L^p(\mathbb{G}^n)} + \left| \int_{\mathbb{R}^n} f \, dV_\gamma \right| \right\} \\
&\le \lim_{p \to \infty} \inf_{f \in \mathcal{A}_\infty(E)} \left\{ \|| \nabla f \|\|_{L^p(\mathbb{G}^n)} + \left| \int_{\mathbb{R}^n} f \, dV_\gamma \right| \right\} \\
&\le \inf_{f \in \mathcal{A}_\infty(E)} \left\{ \|| \nabla f \|\|_{L^\infty(\mathbb{G}^n)} + \left| \int_{\mathbb{R}^n} f \, dV_\gamma \right| \right\} \\
&\simeq \mathrm{Cap}_\infty^{**}(E; \mathbb{G}^n),
\end{aligned}
$$

which proves (5.13).

Now, we show (5.12). Let $p \in [1, \infty)$. Applying

$$
\begin{cases}
\mathcal{A}_\infty(E) \subseteq \mathcal{A}_p(E); \\
\|f\|_{W^{1,p}(\mathbb{G}^n)} \le \|f\|_{W^{1,\infty}(\mathbb{G}^n)} \ \forall \ f \in \mathcal{A}_\infty(E),
\end{cases}
$$

we gain

$$(5.14) \qquad \left(\mathrm{Cap}_p(E; \mathbb{G}^n) \right)^{\frac{1}{p}} \le \inf_{f \in \mathcal{A}_\infty(E)} \|f\|_{W^{1,\infty}(\mathbb{G}^n)} \le \mathrm{Cap}_\infty(E; \mathbb{G}^n).$$

Letting $p \to \infty$ gives

(5.15)
$$\lim_{p \to \infty} \left(\mathrm{Cap}_p(E;\, \mathbb{G}^n) \right)^{\frac{1}{p}} \leq \mathrm{Cap}_\infty(E;\, \mathbb{G}^n).$$

It remains to verify the second inequality of (5.12). For each $p \in [1, \infty)$, choose a function $u_p \in \mathcal{A}_p(E)$ such that

(5.16)
$$\|u_p\|^p_{W^{1,p}(\mathbb{G}^n)} \leq \mathrm{Cap}_p(E;\, \mathbb{G}^n) + 2^{-p^2}.$$

For any $p > q > 1$, we employ (3.6) and (5.14) to get

$$\|u_p\|_{W^{1,q}(\mathbb{G}^n)} \leq 2^{\frac{1}{q}-\frac{1}{p}} \|u_p\|_{W^{1,p}(\mathbb{G}^n)}$$

$$\leq 2^{\frac{1}{q}-\frac{1}{p}} \left(\mathrm{Cap}_p(E;\, \mathbb{G}^n) + 2^{-p^2} \right)^{\frac{1}{p}}$$

$$\leq 2^{\frac{1}{q}-\frac{1}{p}} \mathrm{Cap}_\infty(E;\, \mathbb{G}^n) + 1.$$

Then we apply Proposition 1.3.1 to find a subsequence $\{u_{p_j}\}_{p_j > q}$ (which may depend on $p_{j+1} > p_j > q$) such that

$$\{(u_{p_j}, \nabla u_{p_j})\}_{p_j > q}$$

converges to some

$$(u, \nabla u) \in L^q(\mathbb{G}^n) \times L^q(\mathbb{G}^n; \mathbb{R}^n)$$

in the weak sense.

For any $q_1, q_2 \in (1, \infty)$ such that $q_1 < q_2$, the sequence $\{u_{p_j}\}_{p_j > q_2}$ is bounded in both $W^{1,q_1}(\mathbb{G}^n)$ and $W^{1,q_2}(\mathbb{G}^n)$. Thus, a common subsequence denoted again by $\{u_{p_j}\}_{p_j > q_2}$ can be chosen such that

$$\{(u_{p_j}, \nabla u_{p_j})\}_{p_j > q_2}$$

converges to $(u, \nabla u)$ weakly in both

$$L^{q_1}(\mathbb{G}^n) \times L^{q_1}(\mathbb{G}^n; \mathbb{R}^n)$$

and

$$L^{q_2}(\mathbb{G}^n) \times L^{q_2}(\mathbb{G}^n; \mathbb{R}^n).$$

This implies that there exists a function

$$u \in \cap_{1 < q < \infty} W^{1,q}(\mathbb{G}^n)$$

such that for every $q \in (1, \infty)$, the sequence

$$\{(u_{p_j}, \nabla u_{p_j})\}_{p_j > q}$$

has a subsequence (with indices possibly depending on q) that converges to $(u, \nabla u)$ weakly in

$$L^q(\mathbb{G}^n) \times L^q(\mathbb{G}^n; \mathbb{R}^n).$$

For the moment, we fix some $q \in (n, \infty)$. Applying Mazur's Theorem to the above sequence $\{u_{p_j}\}_{p_j > q}$, we find a sequence $\{v_{p_j}\}_{p_j > q}$ such that every v_{p_j} is a finite convex combination of $\{u_{p_i}\}_{i \geq j}$ and

$$\{(v_{p_j}, \nabla v_{p_j})\}_{p_j > q}$$

converges to $(u, \nabla u)$ strongly in

$$L^q(\mathbb{G}^n) \times L^q(\mathbb{G}^n; \mathbb{R}^n),$$

that is,

$$\lim_{j \to \infty} \left(\|v_{p_j} - u\|_{L^q(\mathbb{G}^n)} + \sum_{i=1}^{n} \left\| \frac{\partial v_{p_j}}{\partial x_i} - \frac{\partial u}{\partial x_i} \right\|_{L^q(\mathbb{G}^n)} \right) = 0.$$

In particular, one has

(5.17)
$$\lim_{j \to \infty} \|v_{p_j} - u\|_{W^{1,q}(\mathbb{G}^n)} = 0.$$

This allows us to extract a subsequence (for simplicity, still denoted by $\{v_{p_j}\}_{p_j > q}$) which converges to u a.e. on $\mathbb{R}^n$, and $\{\nabla v_{p_j}\}_{p_j > q}$ converges to ∇u a.e. on $\mathbb{R}^n$.

Without loss of generality we may assume

$$\begin{cases} v_{p_j} = \sum_{k=j}^{N_j} \lambda_{j,k} u_{p_k}; \\ \lambda_{j,k} \in [0, 1]; \\ j \leq N_j \in \mathbb{N}; \\ \sum_{k=j}^{N_j} \lambda_{j,k} = 1. \end{cases}$$

By the Minkowski inequality, the Hölder inequality, (3.6), and (5.16), we obtain

$$\|v_{p_j}\|_{W^{1,p_j}(\mathbb{G}^n)}^{p_j} = \left\| \sum_{k=j}^{N_j} \lambda_{j,k} u_{p_k} \right\|_{L^{p_j}(\mathbb{G}^n)}^{p_j} + \left\| \sum_{k=j}^{N_j} \lambda_{j,k} \nabla u_{p_k} \right\|_{L^{p_j}(\mathbb{G}^n)}^{p_j}$$

$$\leq \left(\sum_{k=j}^{N_j} \lambda_{j,k} \|u_{p_k}\|_{L^{p_j}(\mathbb{G}^n)} \right)^{p_j} + \left(\sum_{k=j}^{N_j} \lambda_{j,k} \|\nabla u_{p_k}\|_{L^{p_j}(\mathbb{G}^n)} \right)^{p_j}$$

$$\leq \sum_{k=j}^{N_j} \lambda_{j,k} \|u_{p_k}\|_{L^{p_j}(\mathbb{G}^n)}^{p_j} + \sum_{k=j}^{N_j} \lambda_{j,k} \|\nabla u_{p_k}\|_{L^{p_j}(\mathbb{G}^n)}^{p_j}$$

$$= \sum_{k=j}^{N_j} \lambda_{j,k} \|u_{p_k}\|_{W^{1,p_j}(\mathbb{G}^n)}^{p_j}$$

$$\leq \sum_{k=j}^{N_j} \lambda_{j,k} \left(2^{\frac{1}{p_j} - \frac{1}{p_k}} \|u_{p_k}\|_{W^{1,p_k}(\mathbb{G}^n)} \right)^{p_j}$$

$$= \sum_{k=j}^{N_j} \lambda_{j,k} 2^{1-\frac{p_j}{p_k}} \|u_{p_k}\|_{W^{1,p_k}(\mathbb{G}^n)}^{p_j}$$

$$\leq \sum_{k=j}^{N_j} \lambda_{j,k} 2^{1-\frac{p_j}{p_k}} \left(\mathrm{Cap}_{p_k}(E; \mathbb{G}^n) + 2^{-p_k^2} \right)^{\frac{p_j}{p_k}}.$$

Recall that the sequence

$$\left\{ 2^{-\frac{1}{p_k}} \left(\mathrm{Cap}_{p_k}(E; \mathbb{G}^n) \right)^{\frac{1}{p_k}} \right\}_{k=j}^{\infty}$$

increases to

$$\lim_{p \to \infty} \left(\mathrm{Cap}_p(E; \mathbb{G}^n) \right)^{\frac{1}{p}}.$$

Thus, for any

$$k \in \{j, j+1, \ldots, N_j\}$$

one has that $p_k > p_j$ and

$$2^{-\frac{p_j}{p_k}} \left(\mathrm{Cap}_{p_k}(E; \mathbb{G}^n) + 2^{-p_k^2} \right)^{\frac{p_j}{p_k}} \leq 2^{-\frac{p_j}{p_k}} \left(\mathrm{Cap}_{p_k}(E; \mathbb{G}^n) \right)^{\frac{p_j}{p_k}} + 2^{-\frac{p_j}{p_k}} 2^{-p_k p_j}$$

$$\leq \left(\lim_{p \to \infty} \left(\mathrm{Cap}_p(E; \mathbb{G}^n) \right)^{\frac{1}{p}} \right)^{p_j} + 2^{-p_j p_k}$$

$$\leq \left(\lim_{p \to \infty} \left(\mathrm{Cap}_p(E; \mathbb{G}^n) \right)^{\frac{1}{p}} \right)^{p_j} + 2^{-p_j^2},$$

which, plus

$$\sum_{k=j}^{N_j} \lambda_{j,k} = 1,$$

leads to

$$\|v_{p_j}\|_{W^{1,p_j}(\mathbb{G}^n)}^{p_j} \leq 2 \left(\lim_{p \to \infty} \left(\mathrm{Cap}_p(E; \mathbb{G}^n) \right)^{\frac{1}{p}} \right)^{p_j} + 2^{1-p_j^2}.$$

Further, by (5.17), the Fatou lemma and (3.6), we have

$$2^{-\frac{1}{q}} \|u\|_{W^{1,q}(\mathbb{G}^n)} \leq \liminf_{j \to \infty} 2^{-\frac{1}{q}} \|v_{p_j}\|_{W^{1,q}(\mathbb{G}^n)}$$

$$\leq \liminf_{\substack{j \to \infty \\ p_j > q}} 2^{-\frac{1}{p_j}} \|v_{p_j}\|_{W^{1,p_j}(\mathbb{G}^n)}$$

$$\leq \liminf_{\substack{j \to \infty \\ p_j > q}} \left\{ \left(\lim_{p \to \infty} \left(\mathrm{Cap}_p(E; \mathbb{G}^n) \right)^{\frac{1}{p}} \right)^{p_j} + 2^{-p_j^2} \right\}^{\frac{1}{p_j}}$$

$$= \lim_{p \to \infty} \left(\mathrm{Cap}_p(E; \mathbb{G}^n) \right)^{\frac{1}{p}}.$$

Letting $q \to \infty$ in the above inequality and applying (3.5), we deduce

$$
\begin{aligned}
\|u\|_{W^{1,\infty}(\mathbb{G}^n)} &= \|\nabla u\|_{L^\infty(\mathbb{G}^n)} + \|u\|_{L^\infty(\mathbb{G}^n)} \\
&= \lim_{q\to\infty} \left(\|\nabla u\|_{L^q(\mathbb{G}^n)} + \|u\|_{L^q(\mathbb{G}^n)} \right)
\end{aligned}
$$

(5.18)

$$
\le \lim_{q\to\infty} 2^{1-\frac{1}{q}} \left(\|\nabla u\|_{L^q(\mathbb{G}^n)}^q + \|u\|_{L^q(\mathbb{G}^n)}^q \right)^{\frac{1}{q}}
$$

$$
\le 2 \lim_{p\to\infty} \left(\mathrm{Cap}_p(E; \mathbb{G}^n) \right)^{\frac{1}{p}}.
$$

Notice that

$$
u_{p_j} \in \mathcal{A}_{p_j}(E) \Rightarrow E \subseteq \left\{ x \in \mathbb{R}^n : u_{p_j}(x) \ge 1 \right\}^\circ.
$$

So, from the already-proved fact that $\{v_{p_j}\}_{p_j>q}$ converges to u a. e. on $\mathbb{R}^n$, it follows that

$$
E \subseteq \left\{ x \in \mathbb{R}^n : u(x) \ge 1 \right\}
$$

after modifying the function u on a set of measure 0. Since

$$
u \in W^{1,q}(\mathbb{G}^n) \quad \& \quad q \in (n, \infty),
$$

we deduce that u belongs to classical locally Sobolev space $W^{1,q}_{\mathrm{loc}}(\mathbb{R}^n)$, so that u is continuous (see [18, p. 143, Theorem 3(i)]). Consequently, for any $\epsilon \in (0, 1)$,

$$
E \subseteq \left\{ x \in \mathbb{R}^n : u(x) \ge 1 \right\} \subseteq \left\{ x \in \mathbb{R}^n : (1+\epsilon)u(x) > 1 \right\}^\circ.
$$

We therefore obtain

$$
(1+\epsilon)u \in \mathcal{A}_\infty(E)
$$

after modifying u on a set of measure 0.

Finally, (5.18) also tells us that

$$
\mathrm{Cap}_\infty(E; \mathbb{G}^n) \le \|u\|_{W^{1,\infty}(\mathbb{G}^n)} \le 2 \lim_{p\to\infty} \left(\mathrm{Cap}_p(E; \mathbb{G}^n) \right)^{\frac{1}{p}}.
$$

This combined with (5.15) yields (5.12), whence completing the proof of the proposition. □

Remark 5.4.3. Based on the proof of Proposition 5.4.2, the equality between

$$
\mathrm{Cap}_\infty(E; \mathbb{G}^n)
$$

and

$$
\lim_{p\to\infty} \left(\mathrm{Cap}_p(E; \mathbb{G}^n) \right)^{\frac{1}{p}}
$$

still holds whenever in the definition of $\mathrm{Cap}_\infty(E; \mathbb{G}^n)$ the norm $\|f\|_{W^{1,\infty}(\mathbb{G}^n)}$ is replaced by

$$
\max \left\{ \|f\|_{L^\infty(\mathbb{G}^n)}, \|\,|\nabla f|\,\|_{L^\infty(\mathbb{G}^n)} \right\}.
$$

Chapter 6
Gaussian BV-Capacity

Note that the Gaussian perimeter element $dP_\gamma = \gamma dP$ exists as the $(n-1)$-dimensional area element dP with the weight γ. So, $\mathbb{G}^n$ merits a geometric capacity analysis on the functions of bounded variation which are differentiable in the weakest measure theoretic sense. In this chapter we utilize four sections to deal with a Gaussian analogue of the bounded variation capacity of a subset of $\mathbb{R}^n$.

6.1 Basics of $\mathrm{Cap}_{BV}(\cdot; \mathbb{G}^n)$

Through [12] we collect some $\mathbb{G}^n$-groundwork associated with an open subset O of $\mathbb{R}^n$ and its space $C_c^k(O; \mathbb{R}^m)$ of all maps $f : O \to \mathbb{R}^{m \geq 1}$ which have continuous partial derivatives up to order $k \in \{0\} \cup \mathbb{N}$ with compact support.

► The γ-divergence of $v \in C_c^1(O; \mathbb{R}^n)$ is determined by

$$(\mathrm{div}_\gamma v)(x) := \mathrm{div} v(x) - v \cdot x \quad \forall \quad x \in O.$$

► Let $L^p(O; \mathbb{G}^n)$ be the space of all Lebesgue measurable functions f on O with

$$\|f\|_{L^p(O;\mathbb{G}^n)} := \begin{cases} \left(\int_O |f|^p \, dV_\gamma \right)^{\frac{1}{p}} & \text{as} \quad p \in [1, \infty); \\ \inf \{ c : |f(x)| \leq c \text{ for } V_\gamma - \text{a.e. } x \in O \} & \text{as} \quad p = \infty. \end{cases}$$

► The γ-total variation of

$$f \in L^1(O; \mathbb{G}^n)$$

is defined by

$$|Df|_{O;\mathbb{G}^n} := \sup \left\{ \int_O f \, \mathrm{div}_\gamma v \, dV_\gamma : \quad v \in C_c^1(O; \mathbb{R}^n) \text{ with } |v| \leq 1 \right\}.$$

In particular, if $O = \mathbb{R}^n$, then $|Df|_{\mathbb{G}^n}$ stands for $|Df|_{O;\mathbb{G}^n}$, and hence (cf. [27, Theorem 2.7(iii)])

$$|D \max\{f, g\}|_{\mathbb{G}^n} + |D \min\{f, g\}|_{\mathbb{G}^n} \leq |Df|_{\mathbb{G}^n} + |Dg|_{\mathbb{G}^n}.$$

© Springer Nature Switzerland AG 2018
L. Liu et al., *Gaussian Capacity Analysis*, Lecture Notes in Mathematics 2225,
https://doi.org/10.1007/978-3-319-95040-2_6

We say that
$$f \in L^1(O; \mathbb{G}^n)$$
is of the Gaussian bounded variation in O, denoted by
$$f \in BV(O; \mathbb{G}^n),$$
provided
$$|Df|_{O;\mathbb{G}^n} < \infty.$$
When $O = \mathbb{R}^n$, we denote $BV(O; \mathbb{G}^n)$ by $BV(\mathbb{G}^n)$, and write $BV_{loc}(\mathbb{G}^n)$ to express the space of all functions
$$f \in BV(N; \mathbb{G}^n)$$
for each open set $N \subseteq \mathbb{R}^n$ with its closure $\overline{N}$ being compact.

▶ The nature of γ reveals actually that $BV_{loc}(\mathbb{G}^n)$ coincides with $BV_{loc}(\mathbb{R}^n)$ (which is the local space of all Euclidean functions with bounded variation) with a local equivalence of the norms. Consequently, the fine properties of $BV_{loc}(\mathbb{R}^n)$ apply to $BV_{loc}(\mathbb{G}^n)$. Here are two important examples. First, if v_E stands for the outer unit normal to the boundary ∂E of a set E with $P_\gamma(E) < \infty$, then the following Gauss-Green formula:

$$\int_E \text{div}_\gamma v \, dV_\gamma = \int_{\partial^* E} v \cdot v_E \, dP_\gamma \quad \forall \quad v \in C_c^1(\mathbb{R}^n; \mathbb{R}^n)$$

holds with $\partial^* E$ being the reduced boundary of E. Second, for
$$f \in BV(\mathbb{G}^n) \quad \& \quad B(x, r) = \{y \in \mathbb{R}^n : |y - x| < r\}$$
let

$$\begin{cases} \mu_{f;\mathbb{G}^n}(x) := \inf \left\{ t : \lim_{r \to 0} \dfrac{V_\gamma \left(B(x,r) \cap \{y \in \mathbb{R}^n : f(y) > t\} \right)}{V_\gamma \left(B(x,r) \right)} = 0 \right\}; \\[3mm] \lambda_{f;\mathbb{G}^n}(x) := \sup \left\{ t : \lim_{r \to 0} \dfrac{V_\gamma \left(B(x,r) \cap \{y \in \mathbb{R}^n : f(y) < t\} \right)}{V_\gamma \left(B(x,r) \right)} = 0 \right\}. \end{cases}$$

Then (cf. [59, p. 260] or [18, p.216])

$$f(x) = \lim_{r \to 0} \frac{\int_{B(x,r)} f \, dV_\gamma}{V_\gamma \left(B(x,r) \right)} = 2^{-1} \left(\mu_{f;\mathbb{G}^n}(x) + \lambda_{f;\mathbb{G}^n}(x) \right) \text{ for } P_\gamma - a.e. \ x \in \mathbb{R}^n.$$

▶ If f is of the Lipschitz class on $\mathbb{R}^n$, written as $f \in Lip(\mathbb{R}^n)$:
$$|f(x) - f(y)| \leqslant |x - y| \quad \forall \quad (x, y) \in \mathbb{R}^n \times \mathbb{R}^n,$$
then an application of Rademacher's theorem (cf. [18, p.81, Theorem 2]) derives

$$(6.1) \qquad\qquad |Df|_{\mathbb{G}^n} = \int_{\mathbb{R}^n} |\nabla f| \, dV_\gamma < \infty.$$

As a matter of fact, for any

$$v \in C_c^1(\mathbb{R}^n; \mathbb{R}^n) \text{ with } |v| \leq 1$$

an application of the integration-by-parts, the triangle inequality, and the Cauchy-Schwarz inequality derives

$$\left| \int_{\mathbb{R}^n} f \operatorname{div}_\gamma v \, dV_\gamma \right| = \left| \int_{\mathbb{R}^n} v \cdot \nabla f \, dV_\gamma \right| \leq \int_{\mathbb{R}^n} |\nabla f| \, dV_\gamma,$$

whence

$$|Df|_{\mathbb{G}^n} \leq \int_{\mathbb{R}^n} |\nabla f| \, dV_\gamma.$$

To see that this last inequality is actually an equality, just choose a smooth sequence v_j with $|v_j| \leq 1$ such that it approaches $|\nabla f|^{-1} \nabla f$ and note that the equality is valid for v_j.

▶ For a set $E \subseteq \mathbb{R}^n$, let

$$1_E \text{ and } P_\gamma(E) = |D1_E|_{\mathbb{G}^n}$$

be the indicator function and the Gaussian perimeter of E respectively. This definition, plus the above max-min inequality, reveals that the Gaussian perimeter is strongly sub-additive (cf. [12, Lemma 9]):

$$P_\gamma(E_1 \cup E_2) + P_\gamma(E_1 \cap E_2) \leq P_\gamma(E_1) + P_\gamma(E_2) \ \forall \ E_1, E_2 \subseteq \mathbb{R}^n.$$

Moreover, under a smooth condition on the boundary ∂E of E one has

$$P_\gamma(E) = \int_{\partial E} dP_\gamma.$$

▶ As a slight extension of Lemma 5.1.1, the following Gaussian co-area formula holds (cf. [12, Proposition 2] which is from [4, Theorem 3.7]): if

$$f \in BV(\mathbb{G}^n),$$

then

$$|Df|_{\mathbb{G}^n} = \int_{-\infty}^\infty P_\gamma\big(\{x \in \mathbb{R}^n : f(x) > t\}\big) \, dt.$$

Definition 6.1.1. For a set $E \subseteq \mathbb{R}^n$ let $\mathcal{A}(E, BV(\mathbb{G}^n))$ be the class of admissible functions on $\mathbb{G}^n$, i.e., functions

$$f \in BV(\mathbb{G}^n) \text{ with } 0 \leq f \leq 1$$

and $f = 1$ in a neighborhood of E (an open set containing E). The Gaussian BV capacity of E is defined by

$$\mathrm{Cap}_{BV}(E; \mathbb{G}^n) := \inf \left\{ \|f\|_{L^1(\mathbb{G}^n)} + |Df|_{\mathbb{G}^n} : \ f \in \mathcal{A}(E, BV(\mathbb{G}^n)) \right\}.$$

Here it is appropriate to point out the fact that if K is a compact subset of $\mathbb{R}^n$ and $Lip_c(\mathbb{R}^n)$ is the class of all functions

$$f \in Lip(\mathbb{R}^n)$$

with compact support, then

$$\text{Cap}_{BV}(K; \mathbb{G}^n)$$
$$= \inf \left\{ \|f\|_{L^1(\mathbb{G}^n)} + \int_{\mathbb{R}^n} |\nabla f| \, dV_\gamma : f \in Lip_c(\mathbb{R}^n) \cap \mathcal{A}(K, BV(\mathbb{G}^n)) \right\}$$

follows from a regularization argument; see [27, Theorem 3.9] and (6.1). Below is a generalization of Theorem 5.1.2.

Theorem 6.1.2. If E is an arbitrary subset of $\mathbb{R}^n$, then

$$(6.2) \qquad \text{Cap}_{BV}(E; \mathbb{G}^n) = \inf \left\{ V_\gamma(F) + P_\gamma(F) : E \subseteq F^\circ \subseteq \mathbb{R}^n \right\}.$$

Consequently, if K is a compact subset of $\mathbb{R}^n$, then

$$(6.3) \qquad \text{Cap}_{BV}(K; \mathbb{G}^n) = \text{Cap}_{1,*}(K; \mathbb{G}^n) = \text{Cap}_1(K; \mathbb{G}^n).$$

Proof. (6.3) is a straightforward of (6.2) and Theorem 5.1.2. So, it remains to verify (6.2). If $E \subseteq F^\circ$, then

$$1_F \in \mathcal{A}(E, BV(\mathbb{G}^n)),$$

and hence

$$\text{Cap}_{BV}(E; \mathbb{G}^n) \leq \|1_F\|_{L^1(\mathbb{G}^n)} + |D1_F|_{\mathbb{G}^n} = V_\gamma(F) + P_\gamma(F).$$

This leads to one part of (6.2):

$$\text{Cap}_{BV}(E; \mathbb{G}^n) \leq \inf \left\{ V_\gamma(F) + P_\gamma(F) : E \subseteq F^\circ \subseteq \mathbb{R}^n \right\}.$$

Also, if

$$\text{Cap}_{BV}(E; \mathbb{G}^n) < \infty,$$

then Definition 6.1.1 and the Gaussian co-area formula for $BV(\mathbb{G}^n)$ imply that for any $\epsilon > 0$ there is

$$f \in \mathcal{A}(E, BV(\mathbb{G}^n))$$

obeying

$$\int_0^1 \left(V_\gamma(\{x \in \mathbb{R}^n : f(x) > t\}) + P_\gamma(\{x \in \mathbb{R}^n : f(x) > t\}) \right) dt < \text{Cap}_{BV}(E; \mathbb{G}^n) + \epsilon,$$

and hence there exists $t_0 \in (0, 1)$ making

$$V_\gamma(\{x \in \mathbb{R}^n : f(x) > t_0\}) + P_\gamma(\{x \in \mathbb{R}^n : f(x) > t_0\}) < \text{Cap}_{BV}(E; \mathbb{G}^n) + \epsilon.$$

Note that

$$E \subseteq \left\{ x \in \mathbb{R}^n : f(x) > t_0 \right\}^{\circ}.$$

So

$$\text{Cap}_{BV}(E; \mathbb{G}^n) + \epsilon > \inf \left\{ V_\gamma(F) + P_\gamma(F) : E \subseteq F^{\circ} \subseteq \mathbb{R}^n \right\}$$

which, plus letting $\epsilon \to 0$, derives another part of (6.2):

$$\text{Cap}_{BV}(E; \mathbb{G}^n) \geq \inf \left\{ V_\gamma(F) + P_\gamma(F) : E \subseteq F^{\circ} \subseteq \mathbb{R}^n \right\}.$$

$\square$

6.2 Measure Theoretic Nature of $\text{Cap}_{BV}(\cdot; \mathbb{G}^n)$

Theorem 6.2.1. The set-function $\text{Cap}_{BV}(\cdot; \mathbb{G}^n)$ enjoys the following properties.

(i)
$$\text{Cap}_{BV}(\emptyset; \mathbb{G}^n) = 0.$$

(ii)
$$E_1 \subseteq E_2 \subseteq \mathbb{R}^n \Rightarrow \text{Cap}_{BV}(E_1; \mathbb{G}^n) \leq \text{Cap}_{BV}(E_2; \mathbb{G}^n).$$

(iii)
$$\text{Cap}_{BV}\left(\cup_{j=1}^{\infty} E_j; \mathbb{G}^n \right) \leq \sum_{j=1}^{\infty} \text{Cap}_{BV}(E_j; \mathbb{G}^n).$$

(iv)
$$\text{Cap}_{BV}(E_1 \cup E_2; \mathbb{G}^n) + \text{Cap}_{BV}(E_1 \cap E_2; \mathbb{G}^n)$$
$$\leq \text{Cap}_{BV}(E_1; \mathbb{G}^n) + \text{Cap}_{BV}(E_2; \mathbb{G}^n).$$

(v)
$$\lim_{j \to \infty} \text{Cap}_{BV}(E_j; \mathbb{G}^n) = \text{Cap}_{BV}(\cup_{j=1}^{\infty} E_j; \mathbb{G}^n)$$

for any sequence $\{E_j\}_{j=1}^{\infty}$ of subsets of $\mathbb{R}^n$ with $E_1 \subseteq E_2 \subseteq E_3 \subseteq \cdots$.

(vi)
$$\lim_{j \to \infty} \text{Cap}_{BV}(K_j; \mathbb{G}^n) = \text{Cap}_{BV}(\cap_{j=1}^{\infty} K_j; \mathbb{G}^n)$$

for any sequence $\{K_j\}_{j=1}^{\infty}$ of compact subsets of $\mathbb{R}^n$ with $K_1 \supseteq K_2 \supseteq K_3 \supseteq \cdots$.

(vii)
$$\text{Cap}_{BV}(E; \mathbb{G}^n) = \inf \left\{ \text{Cap}_{BV}(O; \mathbb{G}^n) : \text{open } O \supseteq E \right\}.$$

(viii)
$$\text{Cap}_{BV}(E; \mathbb{G}^n) = \sup \left\{ \text{Cap}_{BV}(K; \mathbb{G}^n) : \text{ compact } K \subseteq E \right\}$$

for any Borel set $E \subseteq \mathbb{R}^n$.

Proof. (i)&(ii) follow from Definition 6.1.1.

(iii) Suppose
$$\sum_{j=1}^{\infty} \text{Cap}_{BV}(E_j; \mathbb{G}^n) < \infty.$$

For any $\epsilon > 0$ and each $j = 1, 2, 3, \ldots$ there is
$$f_j \in \mathcal{A}\left(E_j, BV(\mathbb{G}^n)\right)$$

such that
$$\|f_j\|_{L^1(\mathbb{G}^n)} + |Df_j|_{\mathbb{G}^n} < \text{Cap}_{BV}(E_j; \mathbb{G}^n) + \epsilon 2^{-j}.$$

Upon setting
$$f = \sup_j f_j$$

we get
$$f \in \mathcal{A}\left(\cup_{j=1}^{\infty} E_j, BV(\mathbb{G}^n)\right)$$

and
$$\|f\|_{L^1(\mathbb{G}^n)} + |Df|_{\mathbb{G}^n} \leq \sum_{j=1}^{\infty} \|f_j\|_{L^1(\mathbb{G}^n)} + \liminf_{j\to\infty} |D\max\{f_1, \ldots, f_j\}|_{\mathbb{G}^n}$$

$$\leq \sum_{j=1}^{\infty} \left(\|f_j\|_{L^1(\mathbb{G}^n)} + |Df_j|_{\mathbb{G}^n} \right)$$

$$\leq \sum_{j=1}^{\infty} \text{Cap}_{BV}(E_j; \mathbb{G}^n) + \epsilon,$$

thereby reaching the countable-subadditivity through sending ϵ to 0.

(iv) Without losing generality we may assume
$$\text{Cap}_{BV}(E_1; \mathbb{G}^n) + \text{Cap}_{BV}(E_2; \mathbb{G}^n) < \infty.$$

For $\epsilon > 0$ there exist
$$f_1, f_2 \in \mathcal{A}(E_j, BV(\mathbb{G}^n))$$

satisfying
$$\|f_j\|_{L^1(\mathbb{G}^n)} + |Df_j|_{\mathbb{G}^n} < \text{Cap}_{BV}(E_j; \mathbb{G}^n) + 2^{-1}\epsilon \quad \forall \quad j = 1, 2.$$

Evidently, we get
$$\begin{cases} \max\{f_1, f_2\} \in \mathcal{A}(E_1 \cup E_2, BV(\mathbb{G}^n)); \\ \min\{f_1, f_2\} \in \mathcal{A}(E_1 \cap E_2, BV(\mathbb{G}^n)), \end{cases}$$

whence

$$\text{Cap}_{BV}(E_1 \cup E_2; \mathbb{G}^n) + \text{Cap}_{BV}(E_1 \cap E_2; \mathbb{G}^n)$$

$$\leq \|f_1\|_{L^1(\mathbb{G}^n)} + \|f_2\|_{L^1(\mathbb{G}^n)} + |Df_1|_{\mathbb{G}^n} + |Df_2|_{\mathbb{G}^n}$$

$$\leq \sum_{j=1}^{2} \text{Cap}_{BV}(E_j; \mathbb{G}^n) + \epsilon.$$

This, plus letting $\epsilon \to 0$, derives the desired inequality.

(v) Suppose that $\{E_j\}$ is an increasing sequence. On the one hand, we have

$$\lim_{j \to \infty} \text{Cap}_{BV}(E_j; \mathbb{G}^n) \leq \text{Cap}_{BV}(\cup_{j=1}^{\infty} E_j; \mathbb{G}^n).$$

On the other hand, we will verify

$$\lim_{j \to \infty} \text{Cap}_{BV}(E_j; \mathbb{G}^n) \geq \text{Cap}_{BV}(\cup_{j=1}^{\infty} E_j; \mathbb{G}^n).$$

Of course, it suffices to do so assuming

$$\lim_{j \to \infty} \text{Cap}_{BV}(E_j; \mathbb{G}^n) < \infty.$$

For any $\epsilon > 0$ and each $j = 1, 2, 3, \ldots$ there is

$$f_j \in \mathcal{A}(E_j, BV(\mathbb{G}^n))$$

such that

$$\|f_j\|_{L^1(\mathbb{G}^n)} + |Df_j|_{\mathbb{G}^n} < \text{Cap}_{BV}(E_j; \mathbb{G}^n) + \epsilon 2^{-j}.$$

Upon choosing

$$\begin{cases} g_k = \max_{1 \leq j \leq k} f_j; \\ g_0 = 0; \\ E_0 = \emptyset; \\ h_k = \min\{g_{k-1}, f_k\}, \end{cases}$$

we obtain

$$\begin{cases} g_k, h_k \in BV(\mathbb{G}^n); \\ E_{j-1} \subseteq \{x \in \mathbb{R}^n : h_j(x) = 1\}^{\circ}, \end{cases}$$

thereby using $E_j \subseteq E_{j+1}$ to achieve both

$$\|g_k\|_{L^1(\mathbb{G}^n)} + |Dg_k|_{\mathbb{G}^n} + \text{Cap}_{BV}(E_{k-1}; \mathbb{G}^n)$$

$$\leq \|g_k\|_{L^1(\mathbb{G}^n)} + \|h_k\|_{L^1(\mathbb{G}^n)} + |Dg_k|_{\mathbb{G}^n} + |Dh_k|_{\mathbb{G}^n}$$

$$\leq \|g_{k-1}\|_{L^1(\mathbb{G}^n)} + |Dg_{k-1}|_{\mathbb{G}^n} + \text{Cap}_{BV}(E_k; \mathbb{G}^n) + \epsilon 2^{-k}$$

and
$$\|g_j\|_{L^1(\mathbb{G}^n)} + |Dg_j|_{\mathbb{G}^n} \le \mathrm{Cap}_{BV}(E_j; \mathbb{G}^n) + \epsilon.$$

So, if
$$f = \lim_{k\to\infty} g_k,$$

then
$$f \in \mathcal{A}\big(\cup_{j=1}^\infty E_j, BV(\mathbb{G}^n)\big),$$

and hence
$$\begin{aligned}
\mathrm{Cap}_{BV}(\cup_{j=1}^\infty E_j; \mathbb{G}^n) &\le \|f\|_{L^1(\mathbb{G}^n)} + |Df|_{\mathbb{G}^n} \\
&\le \lim_{k\to\infty} \big(\|g_k\|_{L^1(\mathbb{G}^n)} + |Dg_k|_{\mathbb{G}^n}\big) \\
&\le \lim_{k\to\infty} \mathrm{Cap}_{BV}(E_k; \mathbb{G}^n) + \epsilon.
\end{aligned}$$

(vi) Assume that $\{K_j\}_{j=1}^\infty$ is a decreasing sequence of compact subsets of $\mathbb{R}^n$. Then
$$K = \cap_{j=1}^\infty K_j$$

is compact. Referring to the proof of [28, Theorem 2.2(iv)], for any $\epsilon \in (0, 2^{-1})$ there exists
$$f \in \mathcal{A}\big(K, BV(\mathbb{G}^n)\big)$$

such that
$$\|f\|_{L^1(\mathbb{G}^n)} + |Df|_{\mathbb{G}^n} < \mathrm{Cap}_{BV}(K; \mathbb{G}^n) + \epsilon.$$

Note that if j is large enough then K_j is contained in
$$\{x \in \mathbb{R}^n : f(x) \ge 1 - 2^{-1}\epsilon\} \subseteq \{x \in \mathbb{R}^n : f(x) \ge 1 - \epsilon\}^\circ.$$

So
$$\min\{1, (1-\epsilon)^{-1}f\} \in \mathcal{A}(\{x \in \mathbb{R}^n : f(x) \ge 1 - 2^{-1}\epsilon\}, BV(\mathbb{G}^n)).$$

Now an application of (ii) and Definition 6.1.1 yields
$$\begin{aligned}
\lim_{j\to\infty} \mathrm{Cap}_{BV}(K_j; \mathbb{G}^n) &\le \mathrm{Cap}_{BV}(\{x \in \mathbb{R}^n : f(x) \ge 1 - \epsilon\}; \mathbb{G}^n) \\
&\le (1-\epsilon)^{-1}\big(\|f\|_{L^1(\mathbb{G}^n)} + |Df|_{\mathbb{G}^n}\big) \\
&\le (1-\epsilon)^{-1}\big(\mathrm{Cap}_{BV}(K; \mathbb{G}^n) + \epsilon\big).
\end{aligned}$$

Upon sending $\epsilon \to 0$ and employing (ii) again, we achieve the required inequality
$$\mathrm{Cap}_{BV}(K; \mathbb{G}^n) \le \lim_{j\to\infty} \mathrm{Cap}_{BV}(K_j; \mathbb{G}^n) \le \mathrm{Cap}_{BV}(K; \mathbb{G}^n).$$

(vii) According to (ii), we have
$$\mathrm{Cap}_{BV}(E; \mathbb{G}^n) \le \inf\{\mathrm{Cap}_{BV}(O; \mathbb{G}^n) : \text{open } O \supseteq E\}.$$

To reach the reverse inequality, we may assume

$$\text{Cap}_{BV}(E; \mathbb{G}^n) < \infty.$$

Meanwhile, from Definition 6.1.1 it follows that for any $\epsilon > 0$ there exists

$$f \in \mathcal{A}(E, BV(\mathbb{G}^n))$$

satisfying

$$\|f\|_{L^1(\mathbb{G}^n)} + |Df|_{\mathbb{G}^n} < \text{Cap}_{BV}(E; \mathbb{G}^n) + \epsilon.$$

Consequently, there is an open $O \supseteq E$ such that $f = 1$ on O. This in turn derives

$$\text{Cap}_{BV}(O; \mathbb{G}^n) \le \|f\|_{L^1(\mathbb{G}^n)} + |Df|_{\mathbb{G}^n} < \text{Cap}_{BV}(E; \mathbb{G}^n) + \epsilon,$$

and thus

$$\text{Cap}_{BV}(E; \mathbb{G}^n) \ge \inf\{\text{Cap}_{BV}(O; \mathbb{G}^n) : \text{ open } O \supseteq E\}.$$

(viii) This follows from (v) to (vi).   □

Remark 6.2.2. Theorem 6.2.1 reveals that $\text{Cap}_{BV}(\cdot; \mathbb{G}^n)$, as a limit of $\text{Cap}_p(\cdot; \mathbb{G}^n)$ as $p \to 1$, is not only an outer measure or a Meyers capacity (obeying (i)-(ii)-(iii)) (cf. [3]), but also a Choquet capacity (enjoying (i), (ii), (v), (vi)) (cf.[1]). Moreover, a combination of (vii) and (viii) refers to the capacitability.

6.3 Dual Form of $\text{Cap}_{BV}(\cdot; \mathbb{G}^n)$

To describe the dual $[BV(\mathbb{G}^n)]^*$ to $BV(\mathbb{G}^n)$, we refer to [27, Section 5] to introduce the following notion.

Definition 6.3.1. For $E \subseteq \mathbb{R}^n$ and $\epsilon \in (0, \infty)$ let

$$H_{\epsilon, \gamma, n-1}(E) := \inf\left\{\sum_{j=1}^{\infty} r_j^{-1} V_\gamma(B(x_j, r_j)) : E \subseteq \cup_{j=1}^{\infty} B(x_j, r_j) \text{ with } 0 < r_j \le \epsilon\right\}.$$

The quantity

$$H_{\gamma, n-1}(E) := \lim_{\epsilon \to 0} H_{\epsilon, \gamma, n-1}(E)$$

is called the $(n-1)$-dimensional Gaussian Hausdorff measure of E.

Of course, if $V_\gamma(B(x_j, r_j))$ in Definition 6.3.1 is replaced by the Euclidean volume of the ball $B(x_j, r_j)$, then $H_{\gamma, n-1}(E)$ becomes the classical $(n-1)$-dimensional Hausdorff measure $H_{n-1}(E)$. Clearly, if E is a compact subset of $\mathbb{R}^n$, then

$$H_{\gamma, n-1}(E) = 0 \Leftrightarrow H_{n-1}(E) = 0,$$

and hence this equivalence is valid for any Borel subset of $\mathbb{R}^n$.

Lemma 6.3.2. Let $E \subseteq \mathbb{R}^n$ be a Borel set. Then

$$H_{\gamma, n-1}(E) = 0 \Rightarrow \mathrm{Cap}_{BV}(E; \mathbb{G}^n) = 0.$$

Proof. According to Theorem 6.2.1(viii), it suffices to check this last implication for any compact subset K of $\mathbb{R}^n$. Suppose $H_{\gamma, n-1}(K) = 0$. For any $\epsilon \in (0, 1)$ let $\{B(x_j, r_j)\}_{j \in \mathbb{N}}$ be a ball-cover of K with

$$\begin{cases} x_j \in K; \\ r_j \in (0, 1); \\ \sum_{j=1}^{\infty} r_j^{-1} V_\gamma \left(B(x_j, r_j) \right) < \epsilon. \end{cases}$$

Upon taking

$$c = \max_{y \in K} |y| \quad \& \quad f_j(x) = \max \left\{ 0, 1 - \max\{0, r_j^{-1}|x - x_j| - 1\} \right\}$$

and utilizing the definition of $\mathrm{Cap}_{BV}\left(B(x_j, r_j); \mathbb{G}^n \right)$, we estimate

$$\begin{aligned} \mathrm{Cap}_{BV}\left(B(x_j, r_j); \mathbb{G}^n \right) &\leq \|f_j\|_{L^1(\mathbb{G}^n)} + \int_{\mathbb{R}^n} |\nabla f_j| \, dV_\gamma \\ &\leq 2r_j^{-1} \left(V_\gamma \left(B(x_j, r_j) \right) + V_\gamma \left(B(x_j, 2r_j) \right) \right) \\ &\leq 2^{n+2} e^{1+c} r_j^{-1} V_\gamma \left(B(x_j, r_j) \right), \end{aligned}$$

thereby reaching via Theorem 6.2.1(iii)

$$\mathrm{Cap}_{BV}(K; \mathbb{G}^n) \leq \sum_{j=1}^{\infty} \mathrm{Cap}_{BV}\left(B(x_j, r_j); \mathbb{G}^n \right) \leq 2^{n+2} e^{1+c} \epsilon$$

and so

$$\mathrm{Cap}_{BV}(K; \mathbb{G}^n) = 0.$$

$\square$

The coming-up-next assertion characterizes $[BV(\mathbb{G}^n)]^*$ in a similar manner to handle [59, Theorem 5.12.4] on the dual $[BV(\mathbb{R}^n)]^*$ of the Euclidean *BV*-space $BV(\mathbb{R}^n)$.

Theorem 6.3.3. Let μ be a nonnegative Radon measure on $\mathbb{R}^n$. Then the following three statements are equivalent.

(i)

$$\begin{cases} H_{\gamma, n-1}(B) = 0 \Rightarrow \mu(B) = 0 \; \forall \text{ Borel set } B \subseteq \mathbb{R}^n; \\ \left| \int_{\mathbb{R}^n} f \, d\mu \right| \lesssim \|f\|_{L^1(\mathbb{G}^n)} + |Df|_{\mathbb{G}^n} \; \forall \; f \in BV(\mathbb{G}^n). \end{cases}$$

(ii)

$$\mu(B) \lesssim V_\gamma(B) + P_\gamma(B) \quad \forall \quad \text{Borel set} \quad B \subseteq \mathbb{R}^n.$$

(iii)

$$\mu(B) \lesssim \text{Cap}_{BV}(B; \mathbb{G}^n) \quad \forall \quad \text{Borel set} \quad B \subseteq \mathbb{R}^n.$$

Proof. (i)$\Rightarrow$(ii) This follows from the definition of $P_\gamma(B)$.

(ii)$\Rightarrow$(iii) Suppose that (ii) is valid. If K is a compact subset of $\mathbb{R}^n$, then the regularity of μ gives

$$\mu(K) = \inf\{\mu(O) : \text{open } O \supseteq K\} \lesssim \inf\{V_\gamma(O) + P_\gamma(O) : \text{open } O \supseteq K\}.$$

An application of Theorem 6.1.2 derives

$$\mu(K) \lesssim \text{cap}(K, BV(\mathbb{G}^n)).$$

Now, using Theorem 6.2.1(viii) and the inner regularity of μ we get (iii).

(iii)$\Rightarrow$(i) Suppose that (iii) is true. From Lemma 6.3.2 it follows that for any Borel set $B \subseteq \mathbb{R}^n$ then

$$H_{\gamma, n-1}(B) = 0 \Rightarrow \text{Cap}_{BV}(B; \mathbb{G}^n) = 0 \Rightarrow \mu(B) = 0.$$

Notice that

$$H_{\gamma, n-1}(B) = 0 \Leftrightarrow P_\gamma(B) = 0 \quad \forall \quad \text{Borel set} \quad B \subseteq \mathbb{R}^n.$$

So, if $f \in BV(\mathbb{G}^n)$, then f is defined $H_{\gamma, n-1}$-a.e. on $\mathbb{R}^n$, and hence Lemma 6.3.2 yields that any function

$$f \in BV(\mathbb{G}^n)$$

is defined μ-a.e. on $\mathbb{R}^n$. Consequently, if

$$f \in BV(\mathbb{G}^n),$$

then a combination of the layer-cake formula, the assumption (iii), Theorem 6.1.2, and the Gaussian co-area formula for $BV(\mathbb{G}^n)$ derives

$$\left| \int_{\mathbb{R}^n} f \, d\mu \right| \lesssim \int_0^\infty \left(V_\gamma(\{x \in \mathbb{R}^n : |f(x)| > s\}) + P_\gamma(\{x \in \mathbb{R}^n : |f(x)| > s\}) \right) ds$$
$$\lesssim \|f\|_{L^1(\mathbb{G}^n)} + |D|f||_{\mathbb{G}^n}$$
$$\lesssim \|f\|_{L^1(\mathbb{G}^n)} + |Df|_{\mathbb{G}^n},$$

and so (i) follows.   □

Theorem 6.3.3 demonstrates that a nonnegative Radon measure μ obeying any-one of the previous three conditions can be treated as a member of $[BV(\mathbb{G}^n)]^*$ and the induced constant can be chosen as its norm $\|\mu\|_{[BV(\mathbb{G}^n)]^*}$, and moreover, leads to a dual form of $\text{Cap}_{BV}(\cdot; \mathbb{G}^n)$.

Theorem 6.3.4. If $E \subseteq \mathbb{R}^n$ is a Borel set, then

$$\text{Cap}_{BV}(E; \mathbb{G}^n) = \sup_{\mu \in M(\mathbb{G}^n)} \mu(E)$$

where

$$M(\mathbb{G}^n) := \{\text{nonnegative Radon measures } \mu \in [BV(\mathbb{G}^n)]^* \text{ with } \|\mu\|_{[BV(\mathbb{G}^n)]^*} \leq 1\}.$$

Proof. In accordance with Theorem 6.2.1(viii), it suffices to validate the formula for any compact subset of $\mathbb{R}^n$. Given a compact $K \subseteq \mathbb{R}^n$. Suppose:

▸ X is the set of all nonnegative Radon measures μ with support being contained in K and $\mu(\mathbb{R}^n) = 1$;

▸ $\mathcal{Y}$ is the class of all $BV(\mathbb{G}^n)$-functions f with

$$\|f\|_{L^1(\mathbb{G}^n)} + |Df|_{\mathbb{G}^n} \leq 1.$$

Then, X and $\mathcal{Y}$ are convex, X is compact in the weak-star topology, and

$$\mu \mapsto \int_{\mathbb{R}^n} f \, d\mu$$

is lower-semi-continuous on X for each given $f \in \mathcal{Y}$. Clearly,

$$f \in \mathcal{A}(K, BV(\mathbb{G}^n)) \ \& \ \mu \in M(\mathbb{G}^n)$$
$$\Rightarrow \mu(K) \leq \left(\|f\|_{L^1(\mathbb{G}^n)} + |Df|_{\mathbb{G}^n}\right)\|\mu\|_{[BV(\mathbb{G}^n)]^*} \leq \|f\|_{L^1(\mathbb{G}^n)} + |Df|_{\mathbb{G}^n}$$
$$\Rightarrow \mu(K) \leq \text{Cap}_{BV}(K; \mathbb{G}^n)$$
$$\Rightarrow \text{Cap}_{BV}(K; \mathbb{G}^n) \geq \sup_{\mu \in M(\mathbb{G}^n)} \mu(K).$$

To establish the reverse form of the last inequality, we use

$$\sup_{f \in \mathcal{Y}} \int_{\mathbb{R}^n} f \, d\mu = \|\mu\|_{[BV(\mathbb{G}^n)]^*} \Rightarrow \inf_{\mu \in X} \sup_{f \in \mathcal{Y}} \int_{\mathbb{R}^n} f \, d\mu \geq \inf_{\mu \in M(\mathbb{G}^n)} (\mu(K))^{-1}$$

and

$$\inf_{\mu \in X} \int_{\mathbb{R}^n} f \, d\mu = \text{essinf}_{x \in K} f(x) \Rightarrow \sup_{f \in \mathcal{Y}} \inf_{\mu \in X} \int_{\mathbb{R}^n} f \, d\mu \leq \left(\text{Cap}_{BV}(K; \mathbb{G}^n)\right)^{-1}$$

as well as the well-known min-max theorem (cf. [2, Theorem 2.4.1]) to derive the required estimation

$$\left(\sup_{\mu \in M(\mathbb{G}^n)} \mu(K)\right)^{-1} \leq \inf_{\mu \in X} \sup_{f \in \mathcal{Y}} \int_{\mathbb{R}^n} f \, d\mu = \sup_{f \in \mathcal{Y}} \inf_{\mu \in X} \int_{\mathbb{R}^n} f \, d\mu \leq (\text{Cap}_{BV}(K; \mathbb{G}^n))^{-1}.$$

$\square$

6.4 Gaussian BV Isocapacity and Trace Estimation

According to Theorem 5.3.1(ii), the well-known Gaussian isoperimetric inequality says that if B is a Borel subset of $\mathbb{R}^n$, then

$$\Psi\big(V_\gamma(B)\big) \le P_\gamma(B)$$

with equality if B is a half-space of $\mathbb{R}^n$ (cf. [34, (8.15)]), where Ψ is again the Gaussian profile determined by

$$\Psi(t) = \Phi' \circ \Phi^{-1}(t) \quad \& \quad \Phi(t) = (2\pi)^{-\frac{1}{2}} \int_{-\infty}^{t} \exp\left(-\frac{s^2}{2}\right) ds.$$

The following assertion is a slightly extended reformulation of [39, Theorem 1].

Lemma 6.4.1. Let f^* be the nonincreasing rearrangement of f under dV_γ and

$$f^{**}(t) = t^{-1} \int_0^t f^*(s)\, ds.$$

Then the Gaussian isoperimetric inequality is equivalent to any of the following four inequalities.

(i) Bobkov's inequality:

$$\Psi\left(\int_{\mathbb{R}^n} |\nabla f|\, dV_\gamma\right) \le \int_{\mathbb{R}^n} \sqrt{(\Psi(f))^2 + |\nabla f|^2}\, dV_\gamma\ \forall\, f \in Lip_c(\mathbb{R}^n)\, \&\, 0 \le f \le 1.$$

(ii) Ledoux's inequality:

$$\int_0^\infty \Psi\big(V_\gamma(\{x \in \mathbb{R}^n : |f(x)| > s\})\big)\, ds \le \int_{\mathbb{R}^n} |\nabla f|\, dV_\gamma \quad \forall\, f \in Lip_c(\mathbb{R}^n).$$

(iii) Talenti's inequality:

$$(-f^*)'(s)\Psi(s) \le \frac{d}{ds} \int_{\{x \in \mathbb{R}^n : |f(x)| > f^*(s)\}} |\nabla f|\, dV_\gamma \quad \forall\, f \in Lip_c(\mathbb{R}^n).$$

(iv) Martín-Milman's oscillation inequality:

$$\big(f^{**}(t) - f^*(t)\big)\Psi(t) \le t|\nabla f|^{**}(t) \quad \forall\, f \in Lip_c(\mathbb{R}^n).$$

Proof. According to [33], Bobkow's inequality evidently implies the Gaussian isoperimetric inequality via approximating the indicator function 1_B by $Lip_c(\mathbb{R}^n)$-functions. Conversely, if the Gaussian isoperimetric inequality acts on

$$B = \big\{(x, t) \in \mathbb{R}^n \times \mathbb{R}^n : \Phi(t) < f(x)\big\} \subseteq \mathbb{R}^{n+1},$$

then Bobkow's inequality follows right away.

Moreover, as shown in [6] Bobkow's inequality can be verified by the basic inequality

$$\Phi\left(\frac{a+b}{2}\right) \leq 2^{-1}\left(\sqrt{(\Phi(a))^2 + \left(\frac{a-b}{2}\right)^2} + \sqrt{(\Phi(b))^2 + \left(\frac{a-b}{2}\right)^2}\right) \quad \forall\, a,b \in [0,1].$$

Here, it is appropriate to mention that (cf. [11, Theorem 1]) if a Lebesgue measurable function f on $\mathbb{R}^n$ obeys

$$\Psi\left(\int_{\mathbb{R}^n} |\nabla f|\, dV_\gamma\right) = \int_{\mathbb{R}^n} \sqrt{(\Psi(f))^2 + |\nabla f|^2}\, dV_\gamma,$$

then either

$$f(x) = \Phi(\ell \cdot x + b)$$

holds for some

$$(\ell, b) \in \mathbb{R}^n \times \mathbb{R} \quad \text{or} \quad f = 1_H$$

of the half-space

$$H = \{x \in \mathbb{R}^n : x \cdot \ell < b\}$$

for some $(\ell, b) \in \mathbb{R}^n \times \mathbb{R}$. □

The foregoing lemma can be utilized to reveal the following equivalence.

Theorem 6.4.2. The Gaussian isoperimetric inequality is equivalent to the Gaussian BV isocapacitary inequality

$$V_\gamma(B) + \Psi(V_\gamma(B)) \leq \mathrm{Cap}_{BV}(B; \mathbb{G}^n) \quad \forall \quad \text{Borel set} \quad B \subseteq \mathbb{R}^n$$

with equality if B is a half-space of $\mathbb{R}^n$.

Proof. Suppose that the Gaussian isoperimetric inequality is valid. Then an application of Lemma 6.4.1 ensures the truth of Martín-Milman's oscillation inequality. Given a compact set $K \subseteq \mathbb{R}^n$. Using Definition 6.1.1 we can get a sequence

$$\{f_j\} \subseteq Lip_c(\mathbb{R}^n) \cap \mathcal{A}(K, BV(\mathbb{G}^n))$$

such that f_j tends to 1_K in $L^1(\mathbb{G}^n)$ and

$$\mathrm{Cap}_{BV}(K; \mathbb{G}^n) = V_\gamma(K) + \liminf_{j\to\infty} \int_{\mathbb{R}^n} |\nabla f_j|\, dV_\gamma,$$

thereby reaching via Lemma 6.4.1(iv)

$$\liminf_{j\to\infty} \Psi(t)(f_j^{**}(t) - f_j^*(t)) \leq \liminf_{j\to\infty} \int_{\mathbb{R}^n} |\nabla f_j|\, dV_\gamma.$$

By virtue of the argument for the final part of [39, p. 162, (iv)⇒(i)], we get

$$V_\gamma(K) + \Psi(V_\gamma(K)) \leq \mathrm{Cap}_{BV}(K; \mathbb{G}^n).$$

An application of the regularity of V_γ and Theorem 6.2.1(viii) deduces that the last Gaussian BV isocapacitary inequality is valid for any Borel set $B \subseteq \mathbb{R}^n$.

Meanwhile, if the previously-stated Gaussian BV isocapacitary inequality is true for any Borel set $B \subseteq \mathbb{R}^n$, then an application of Theorem 6.1.2 derives

$$V_\gamma(B) + \Psi\big(V_\gamma(B)\big) \leq V_\gamma(B) + P_\gamma(B) \quad \forall \quad \text{Borel set} \quad B \subseteq \mathbb{R}^n,$$

and hence, the Gaussian isoperimetric inequality is true for any Borel set $B \subseteq \mathbb{R}^n$.

Note that if B is a half-space of $\mathbb{R}^n$, then equality of the Gaussian isoperimetric inequality holds. So an application of the last estimate derives that equality of the Gaussian isocapacitary inequality must be true for the half-space B. □

According to [34, p. 272], if

$$f \in C^1(\mathbb{R}^n),$$

then

$$\int_{\mathbb{R}^n} |\nabla f| \, dV_\gamma < \infty \Rightarrow \int_{\mathbb{R}^n} |f| \sqrt{\ln(1 + |f|)} \, dV_\gamma < \infty.$$

This induces a trace assertion for the Gauss-Sobolev 1-space of all $C_c^\infty(\mathbb{R}^n)$-functions f with norm $\big\|\,|\nabla f|\,\big\|_{L^1(\mathbb{G}^n)}$ as follows.

Theorem 6.4.3. Let μ be a nonnegative Radon measure on $\mathbb{R}^n$. Then

$$(6.4) \qquad \int_{\mathbb{R}^n} |f| \sqrt{\ln(1 + |f|)} \, d\mu \lesssim \int_{\mathbb{R}^n} |\nabla f| \, dV_\gamma \quad \forall \quad f \in C_c^1(\mathbb{R}^n)$$

if and only if

$$(6.5) \qquad \mu(O) \lesssim P_\gamma(O) \quad \forall \quad O \in \mathcal{O}^n$$

where $\mathcal{O}^n$ is the family of all open sets $O \subseteq \mathbb{R}^n$ with $\overline{O}$ and ∂O being compact and of C^1 respectively.

Proof. For $O \in \mathcal{O}^n$ let $\text{dist}(x, O)$ be the Euclidean distance from x to O and

$$O_t = \{x \in \mathbb{R}^n : \text{dist}(x, O) < t\} \quad \forall \quad t > 0.$$

Given $0 < \epsilon < 1$. Choosing a decreasing C^∞-function $\varphi : [0, 1] \to \mathbb{R}$ such that

$$\varphi(0) = 1 \quad \& \quad \varphi(t) = 0 \quad \forall \quad t > \epsilon,$$

we obtain that if

$$f_O(x) = \varphi\big(\text{dist}(x, O)\big)$$

then an application of the Gaussian co-area formula for $BV(\mathbb{G}^n)$ produces

$$\int_{\mathbb{R}^n} |\nabla f_O| \, dV_\gamma = -\int_0^\epsilon P_\gamma(O_t) \varphi'(t) \, dt \to P_\gamma(O) \quad \text{as} \quad \epsilon \to 0.$$

Note that

$$\int_{\mathbb{R}^n} |f_O|\sqrt{\ln(1+|f_O|)}\, d\mu \geq \int_O |\varphi(\text{dist}(x,O))|\sqrt{\ln\left(1+|\varphi(\text{dist}(x,O))|\right)}\, d\mu(x)$$

$$\gtrsim \mu(O).$$

Thus, upon assuming (6.4) we use the test function f_O to validate (6.5).

Suppose that (6.5) holds. Let

$$\begin{cases} t \in [0,\infty); \\ f \in C_c^1(\mathbb{R}^n); \\ O_f(t) = \{x \in \mathbb{R}^n : |f(x)| > t\}; \\ \phi(t) = t\sqrt{\ln(1+t)}; \\ \psi(t) = \sqrt{\ln(1+t)}. \end{cases}$$

An efficient modification of the standard truncation method (cf. [41, p.155, Remark 1]) leads to a consideration of

$$\begin{cases} \tau \in C^\infty(\mathbb{R}); \\ 0 \leq \tau'(t) \leq 2 \quad \forall \quad t > 0; \\ \tau(t) = \begin{cases} 1 & \text{as } t \geq 1 \\ 0 & \text{as } t \leq 0; \end{cases} \\ f_k(x) = \tau\left(\psi(|f(x)|) - \psi(2^k) + 1\right) \quad \forall \quad (k,x) \in \mathbb{Z} \times \mathbb{R}^n. \end{cases}$$

Note that

$$\psi(2^{k-1}) \leq \psi(2^k) \leq 1 + \psi(2^{k-1}) \quad \forall \quad k \in \mathbb{Z} \Rightarrow 0 \leq f_k \leq 1 \quad \& \quad f_k|_{O_f(2^k)} = 1.$$

So a combination of the layer-cake formula and Theorem 6.2.1(ii) derives

$$\int_{\mathbb{R}^n} |f|\sqrt{\ln(1+|f|)}\, d\mu$$

$$= \int_0^\infty \mu(O_f(t))\, d\phi(t)$$

$$= \sum_{k \in \mathbb{Z}} \int_{2^k}^{2^{k+1}} \mu(O_f(t))\, d\phi(t)$$

$$\leq \left(\sup_{O \in \mathcal{O}^n} \frac{\mu(O)}{P_\gamma(O)}\right) \sum_{k \in \mathbb{Z}} \int_{2^k}^{2^{k+1}} P_\gamma(O_f(t))\, d\phi(t)$$

$$\leq \left(\sup_{O \in \mathcal{O}^n} \frac{\mu(O)}{P_\gamma(O)}\right) \sum_{k \in \mathbb{Z}} \phi(2^{k+1}) \int_{\mathbb{R}^n} |\nabla f_k|\, dV_\gamma$$

$$\leq \left(\sup_{O \in \mathcal{O}^n} \frac{\mu(O)}{P_\gamma(O)}\right) \sum_{k \in \mathbb{Z}} \phi(2^{k+1}) \int_{O_f(2^k)\backslash O_f(2^{k-1})} \frac{|\nabla f|}{(1+|f|)\sqrt{\ln(1+|f|)}}\, dV_\gamma$$

$$\leq \left(\sup_{O \in \hat{O}^n} \frac{\mu(O)}{P_\gamma(O)} \right) \sum_{k \in \mathbb{Z}} \left(\frac{\phi(2^{k+1})}{\phi(2^{k-1})} \right) \int_{O_f(2^k) \setminus O_f(2^{k-1})} |\nabla f| \, dV_\gamma$$

$$\leq \left(\sup_{O \in \hat{O}^n} \frac{\mu(O)}{P_\gamma(O)} \right) \int_{\mathbb{R}^n} |\nabla f| \, dV_\gamma,$$

thereby verifying (6.4).   □

Motivated by Lemmas 6.3.3 & 6.4.1 and the following optimal Gauss-Poincaré inequality (cf. Theorem 5.2.1 or [34, p.115] and its reference [56])

$$\int_{\mathbb{R}^n} \left| f - \int_{\mathbb{R}^n} f \, dV_\gamma \right| dV_\gamma \leq \sqrt{\frac{\pi}{2}} \int_{\mathbb{R}^n} |\nabla f| \, dV_\gamma \quad \forall \quad f \in C_c^1(\mathbb{R}^n),$$

we discover the following $BV(\mathbb{G}^n)$ trace result which is analogous to the classical one for $BV(\mathbb{R}^n)$ presented in [59, Theorem 5.13.1].

Theorem 6.4.4. Let $1 \leq p < \infty$ and μ be a nonnegative Radon measure on $\mathbb{R}^n$. Then the following are equivalent.

(i)

$$\begin{cases} H_{\gamma, n-1}(B) = 0 \Rightarrow \mu(B) = 0 \quad \forall \quad \text{Borel set } B \subseteq \mathbb{R}^n; \\ \left(\int_{\mathbb{R}^n} |f|^p \, d\mu \right)^{\frac{1}{p}} \lesssim \|f\|_{L^1(\mathbb{G}^n)} + |Df|_{\mathbb{G}^n} \ \forall f \in BV(\mathbb{G}^n). \end{cases}$$

(ii)

$$\mu(B) \lesssim \left(V_\gamma(B) + P_\gamma(B) \right)^p \quad \forall \quad \text{Borel set} \quad B \subseteq \mathbb{R}^n.$$

(iii)

$$\mu(B) \lesssim \left(\text{Cap}_{BV}(B; \mathbb{G}^n) \right)^p \quad \forall \quad \text{Borel set} \quad B \subseteq \mathbb{R}^n.$$

Moreover

$$\sup_{f \in C_c^1(\mathbb{R}^n)} \frac{\left(\int_{\mathbb{R}^n} |f|^p \, d\mu \right)^{\frac{1}{p}}}{\|f\|_{L^1(\mathbb{G}^n)} + \int_{\mathbb{R}^n} |\nabla f| \, dV_\gamma} < \infty \Leftrightarrow \sup_{O \in \hat{O}^n} \frac{(\mu(O))^{\frac{1}{p}}}{\text{Cap}_{BV}(O; \mathbb{G}^n)} < \infty.$$

Proof. (i)$\Rightarrow$(ii) This follows from taking a function $f = 1_B$ for any Borel set $B \subseteq \mathbb{R}^n$.

(ii)$\Rightarrow$(iii) Suppose that (ii) is true. If K is a compact subset of $\mathbb{R}^n$, then the regularity of μ ensures

$$\mu(K) = \inf \{ \mu(O) : \forall \text{ open } O \supseteq K \} \lesssim \inf \{ V_\gamma(O) + P_\gamma(O) : \forall \text{ open } O \supseteq K \}.$$

Using Theorem 6.1.2 we get

$$(\mu(K))^{\frac{1}{p}} \lesssim \text{Cap}_{BV}(K; \mathbb{G}^n),$$

whence (iii) by virtue of Theorem 6.2.1(viii) and the inner regularity of μ.

(iii)$\Rightarrow$(i) Assume that (iii) holds. By Lemma 6.3.2 we have that if

$$H_{\gamma,n-1}(B) = 0 \quad \forall \quad \text{Borel set} \quad B \subseteq \mathbb{R}^n$$

then

$$\text{Cap}_{BV}(B; \mathbb{G}^n) = 0 \quad \forall \quad \text{Borel set} \quad B \subseteq \mathbb{R}^n,$$

thereby finding via the assumption (iii)

$$\mu(B) = 0 \quad \forall \quad \text{Borel set} \quad B \subseteq \mathbb{R}^n.$$

Since

$$f \in BV(\mathbb{G}^n) \text{ is defined } H_{\gamma,n-1} - \text{a.e. on } \mathbb{R}^n,$$

we see that $|f|^p$ is defined μ-a.e. on $\mathbb{R}^n$.

Also, if

$$0 \le g \quad \& \quad \infty > \|g\|_{L^{\frac{p}{p-1}}(\mathbb{R}^n;\mu)} = \begin{cases} \left(\int_{\mathbb{R}^n} g^{\frac{p}{p-1}}\, d\mu\right)^{\frac{p-1}{p}} & \text{as } p \in (1,\infty); \\ \sup\left\{g(x): \mu - \text{a.e. } x \in \mathbb{R}^n\right\} & \text{as } p = 1, \end{cases}$$

then the Hölder inequality yields

$$\int_B g\, d\mu \lesssim \|g\|_{L^{\frac{p}{p-1}}(\mathbb{R}^n;\mu)} \text{Cap}_{BV}(B; \mathbb{G}^n) \quad \forall \quad \text{Borel set} \quad B \subseteq \mathbb{R}^n,$$

and hence

$$\nu(B) = \int_B g\, d\mu \quad \forall \quad \text{Borel set} \quad B \subseteq \mathbb{R}^n$$

is well defined and enjoys Theorem 6.4.1(iii). This in turn derives

$$\left|\int_{\mathbb{R}^n} fg\, d\mu\right| \lesssim \|g\|_{L^{\frac{p}{p-1}}(\mathbb{R}^n;\mu)} \left(\|f\|_{L^1(\mathbb{G}^n)} + |Df|_{\mathbb{G}^n}\right) \quad \forall \quad f \in BV(\mathbb{G}^n),$$

thereby revealing

$$\left(\int_{\mathbb{R}^n} |f|^p\, d\mu\right)^{\frac{1}{p}} \lesssim \|f\|_{L^1(\mathbb{G}^n)} + |Df|_{\mathbb{G}^n} \quad \forall \quad f \in BV(\mathbb{G}^n)$$

through the duality

$$[L^p(\mathbb{R}^n;\mu)]^* = L^{\frac{p}{p-1}}(\mathbb{R}^n;\mu).$$

Next, let us check the equivalence after "Moreover" in Theorem 6.4.4. If

$$(6.6) \qquad \sup_{f \in C_c^\infty(\mathbb{R}^n)} \frac{\left(\int_{\mathbb{R}^n} |f|^p\, d\mu\right)^{\frac{1}{p}}}{\|f\|_{L^1(\mathbb{G}^n)} + \int_{\mathbb{R}^n} |\nabla f|\, dV_\gamma} < \infty,$$

then an application of the test function f_O used in the argument for the "only if" part of Theorem 6.4.3 readily implies

(6.7)
$$\sup_{O \in \mathcal{O}^n} \frac{(\mu(O))^{\frac{1}{p}}}{\mathrm{Cap}_{BV}(O; \mathbb{G}^n)} < \infty.$$

Conversely, let (6.7) hold. Using the layer-cake formula, Theorem 6.1.2, and the Gaussian co-area formula for $BV(\mathbb{G}^n)$ we obtain (6.6) via the implication that if

$$f \in C_c^1(\mathbb{R}^n),$$

then

$$\left(\int_{\mathbb{R}^n} |f|^p \, d\mu \right)^{\frac{1}{p}}$$

$$\leq \left(\int_0^\infty \mu(\{x \in \mathbb{R}^n : |f(x)| > t\}) \, dt^p \right)^{\frac{1}{p}}$$

$$= \left(\int_0^\infty \frac{d}{dt} \left(\int_0^t \mu(\{x \in \mathbb{R}^n : |f(x)| > s\}) \, ds^p \right) dt \right)^{\frac{1}{p}}$$

$$= \int_0^\infty \frac{d}{dt} \left(\int_0^t \mu(\{x \in \mathbb{R}^n : |f(x)| > s\}) \, ds^p \right)^{\frac{1}{p}} dt$$

$$\leq \int_0^\infty \left(\mu(\{x \in \mathbb{R}^n : |f(x)| > t\}) \right)^{\frac{1}{p}} dt$$

$$\leq \left(\sup_{O \in \mathcal{O}^n} \frac{(\mu(O))^{\frac{1}{p}}}{\mathrm{Cap}_{BV}(O; \mathbb{G}^n)} \right) \int_0^\infty \mathrm{Cap}_{BV}(\{x \in \mathbb{R}^n : |f(x)| > t\}; \mathbb{G}^n) \, dt$$

$$\leq \left(\sup_{O \in \mathcal{O}^n} \frac{(\mu(O))^{\frac{1}{p}}}{\mathrm{Cap}_{BV}(O; \mathbb{G}^n)} \right) \left(\|f\|_{L^1(\mathbb{G}^n)} + |D|f|\big|_{\mathbb{G}^n} \right)$$

$$\leq \left(\sup_{O \in \mathcal{O}^n} \frac{(\mu(O))^{\frac{1}{p}}}{\mathrm{Cap}_{BV}(O; \mathbb{G}^n)} \right) \left(\|f\|_{L^1(\mathbb{G}^n)} + \int_{\mathbb{R}^n} |\nabla f| \, dV_\gamma \right).$$

$\square$

Bibliography

[1] D.R. Adams, Choquet integrals in potential theory. Publ. Math. **42**, 3–66 (1998)

[2] D.R. Adams, L.I. Hedberg, *Function Spaces and Potential Theory* (Springer, Berlin/Heidelberg, 1996)

[3] D.R. Adams, J. Xiao, Morrey spaces in harmonic analysis. Ark. Mat. **50**, 201–230 (2012)

[4] L. Ambrosio, M. Miranda Jr, S. Maniglia, D. Pallara, BV functions in abstract Wiener spaces. J. Funct. Anal. **258**, 785–813 (2010)

[5] S. Artstein-Avidan, Y. Ostrover, A Brunn-Minkowski inequality for symplectic capacities of convex domains. Int. Math. Res. Not. IMRN 2008, no. 13, Art. ID rnn044, 31 pp. (2008)

[6] S.G. Bobkov, An isoperimetric inequality on the discrete cube, and an elementary proof of the isoperimetric inequality in Gauss space. Ann. Probab. **25**, 206–214 (1997)

[7] H. Boedihardjo, X. Geng, Z. Qian, Quasi-sure existence of Gaussian rough paths and large deviation principles for capacities. Osaka J. Math. **53**, 941–970 (2016)

[8] V.I. Bogachev, *Gaussian Measures*. Mathematical Surveys and Monographs, vol. 62 (American Mathematical Society, Providence, 1998)

[9] C. Borell, The Ehrhard inequality. C.R. Acad. Sci. Paris Ser. I **337**, 663–666 (2003)

[10] B. Brandolini, F. Chiacchio, A. Henrot, C. Trombetti, An optimal Poincaré-Wirtinger inequality in Gauss space. Math. Res. Lett. **20**, 449–457 (2013)

[11] E.A. Carlen, C. Kerce, On the cases of equality in Bobkov's inequality and Gaussian rearrangement. Calc. Var. Partial Differ. Equ. **13**, 1–18 (2001)

[12] V. Caselles, M. Miranda Jr, M. Novaga, Total variation and Cheeger sets in Gauss space. J. Funct. Anal. **259**, 1491–1516 (2010)

[13] J. Cheeger, A lower bound for the smallest eigenvalue of the Laplacian, in *Problems in Analysis* (Papers dedicated to Salomon Bochner, 1969), pp. 195–199 (Princeton University Press, Princeton, NJ, 1970)

[14] G. Choquet, Theory of capacities. Ann. Inst. Fourier Grenoble **5**, 131–295 (1953/1954)

© Springer Nature Switzerland AG 2018
L. Liu et al., *Gaussian Capacity Analysis*, Lecture Notes in Mathematics 2225,
https://doi.org/10.1007/978-3-319-95040-2

[15] S. Costea, Strong A_∞-weights and scaling invariant Besov capacities. Rev. Mat. Iberoam. **23**, 1067–1114 (2007)

[16] S. Costea, Sobolev capacity and Hausdorff measures in metric measure spaces. Ann. Acad. Sci. Fenn. Math. **34**, 179–194 (2009)

[17] A. Ehrhard, Symétrisation dans l'espace de Gauss. Math. Scand. **53**, 281–301 (1983)

[18] L.C. Evans, R.F. Gariepy, *Measure Theory and Fine Properties of Functions*. Studies in Advanced Mathematics (CRC, Boca Raton, 1992)

[19] S. Fang, V. Nolot, Sobolev estimates for optimal transport maps on Gaussian space. J. Funct. Anal. **266**, 5045–5084 (2014)

[20] M. Focardi, M.S. Gelli, G. Pisante, On a 1-capacitary type problem in the plane. Commun. Pure Appl. Anal. **9**, 1319–1333 (2010)

[21] M. Fukushima, Y. Oshima, M. Takeda, *Dirichlet Forms and Symmetric Markov Processes*. De Gruyter Studies in Mathematics, vol. 19 (Walter de Gruyter & Co., Berlin, 1994)

[22] M. Fukushima, T. Uemura, On Sobolev and capacitary inequalities for contractive Besov spaces over d-sets. Potential Anal. **18**, 59–77 (2003)

[23] A.E. Gatto, W. Urbina, On Gaussian Lipschitz spaces and the boundedness of fractional integrals and fractional derivatives on them. Quaest. Math. **38**, 1–25 (2015)

[24] A.E. Gatto, E. Pineda, W. Urbina, Riesz potentials, Bessel potentials, and fractional derivatives on Besov-Lipschitz spaces for the Gaussian measure, in *Recent Advances in Harmonic Analysis and Applications*. Springer Proceedings in Mathematics and Statistics, vol. 25 (Springer, New York, 2013), pp. 105–130

[25] I. Gentil, Logarithmic Sobolev inequality for diffusion semigroups, in *Optimal Transportation*. London Mathematical Society Lecture Note Series, vol. 413 (Cambridge University Press, Cambridge, 2014), pp. 41–57

[26] M. Goldman, M. Novaga, Approximation and relaxation of perimeter in the Wiener space. Ann. Inst. H. Poincaré Anal. Non Linéaire **29**, 525–544 (2012)

[27] H. Hakkarainen, J. Kinnunen, The BV-capacity in metric spaces. Manuscripta Math. **132**, 51–73 (2010)

[28] J. Heinonen, T. Kilpeläinen, O. Martio, *Nonlinear Potential Theory of Degenerate Elliptic Equations* (Oxford University Press, New York, 1993)

[29] N. Jacob, R.L. Schilling, Towards an L^p potential theory for sub-Markovian semigroups: kernels and capacities. Acta Math. Sin. (Engl. Ser.) **22**, 1227–1250 (2006)

[30] J. Kinnunen, O. Martio, The Sobolev capacity on metric spaces. Ann. Acad. Sci. Fenn. Math. **21**, 367–382 (1996)

[31] J. Kinnunen, O. Martio, Choquet property for the Sobolev capacity in metric spaces. in *Proceedings on Analysis and Geometry (Russian) (Novosibirsk Akademgorodok, 1999), 285–290* (Izdat. Ross. Akad. Nauk Sib. Otd. Inst. Mat., Novosibirsk, 2000)

[32] R. Latala, A note on the Ehrhard inequality. Stud. Math. **118**, 169–174 (1996)

[33] R. Latala, On some inequalities for Gaussian measures. Proc. ICM **2**, 813–822 (2002)

[34] M. Ledoux, Isoperimetry and Gaussian analysis. Lect. Notes Math. **1648**, 165–294 (1996)

[35] L. Liu, P. Sjögren, A characterization of the Gaussian Lipschitz space and sharp estimates for the Ornstein-Uhlenbeck Poisson kernel. Rev. Mat. Iberoam. **32**, 1189–1210 (2016)

[36] L. Liu, Y. Sawano, D. Yang, Morrey-type spaces on Gauss measure spaces and boundedness of singular integrals. J. Geom. Anal. **24**, 1007–1051 (2014)

[37] L. Liu, D. Yang, Pointwise multipliers for Campanato spaces on Gauss measure spaces. Nagoya Math. J. **214**, 169–193 (2014)

[38] P. Malliavin, *Stochastic Analysis* (Springer, Berlin, 1997)

[39] J. Martin, M. Milman, Isoperimetry and symmetrization for logarithmic Sobolev inequalities. J. Funct. Anal. **256**, 149–178 (2009)

[40] G. Mauceri, S. Meda, *BMO* and H^1 for the Ornstein-Uhlenbeck operator. J. Funct. Anal. **252**, 278–313 (2007)

[41] V. Maz'ya, *Sobolev Spaces with Applications to Elliptic Partial Differential Equations*, 2nd, revised and augmented edn. (Springer, Heidelberg, 2011), xxviii+866 pp.

[42] A. Messiah, *Quantum Mechanics*, vol. I (North-Holland, Amsterdam, 1999)

[43] F. Morgan, Manifolds with density. Not. Am. Math. Soc. **52**, 853–858 (2005)

[44] J.M. Pearson, The Poincaré inequality and entire functions. Proc. Am. Math. Soc. **118**, 1193–1197 (1993)

[45] J.M. Pearson, An elementary proof of Gross' logarithmic Sobolev inequality. Bull. Lond. Math. Soc. **25**, 463–466 (1993)

[46] E. Pineda, W. Urbina, Some results on Gaussian Besov-Lipschitz spaces and Gaussian Triebel-Lizorkin spaces. J. Approx. Theory **161**, 529–564 (2009)

[47] G. Pisier, Probabilistic methods in the geometry of Banach spaces, in *Probability and Analysis (Varenna, 1985), 167–241.* Lecture Notes in Mathematics, vol. 1206 (Springer, Berlin, 1986)

[48] G. Pólya, Estimating electrostatic capacity. Am. Math. Mon. **54**, 201–206 (1947)

[49] W. Rudin, *Functional Analysis* (McGraw-Hill, New York, 1991)

[50] M. Shubin, Capacity and its applications. http://citeseerx.ist.psu.edu/viewdoc/down. https://doi.org/10.1.1.140.1404&rep=rep1&type=pdf

[51] P. Sjögren, Operators associated with the Hermite semigroup - a survey. J. Fourier Anal. Appl. **3**, 813–823 (1997)

[52] E.M. Stein, *Singular Integrals and Differentiability Properties of Functions* (Princeton University Press, Princeton, NJ, 1970), xiv+290 pp.

[53] J. Xiao, Gaussian *BV*-capacity. Adv. Calc. Var. **9**, 187–200 (2016)

[54] J. Xiao, N. Zhang, Isocapacity estimates for Hessian operators. J. Funct. Anal. **267**, 579–604 (2014)

[55] J. Xiao, N. Zhang, The relative *p*-affine capacity. Proc. Am. Math. Soc. **144**, 3537–3554 (2016)

[56] S.-T. Yau, Isoperimetric constants and the first eigenvalue of a compact Riemannian manifold. Ann. Scient. Éc. Norm. Sup. **8**, 487–507 (1975)

[57] K. Yosida, *Functional Analysis.* 6th edn. (Springer, Berlin/New York, 1980)

[58] Q. Zeng, Poincaré type inequalities for group measure spaces and related transportation cost inequalities, J. Funct. Anal. **266**, 3236–3264 (2014)

[59] W.P. Ziemer, *Weakly Differentiable Functions* (Springer, New York, 1989)

Index

Symbols

1_E, vii
$\mathcal{A}_\infty(E)$, 77
$\mathcal{A}_p(E)$, 37
$\mathcal{A}(E, BV(\mathbb{G}^n))$, 85
$\mathcal{A}(K)$, 48
$\mathcal{B}_a$, 19
$[BV(\mathbb{G}^n)]^*$, 92
$BV(O; \mathbb{G}^n)$, 84
$BV(\mathbb{G}^n)$, 84
$BV_{\text{loc}}(\mathbb{G}^n)$, 84
$C_{\mathcal{B}_a}^{p,\kappa}(\mathbb{G}^n)$, 20
$C_c^0(\mathbb{R}^n)$, v
$C_c^0(\mathbb{R}^n; \mathbb{R}^n)$, 37
$C_c^1(\mathbb{R}^n; \mathbb{R}^n)$, 37
$C^k(\mathbb{R}^n)$, vii
$C_c^k(\mathbb{R}^n)$, vii
$C_c^k(\mathbb{R}^n)$, v
$\text{Cap}_{0,p}(\cdot; \mathbb{G}^n)$, 48, 50
$\text{Cap}_{1,*}(\cdot; \mathbb{G}^n)$, 66
$\text{Cap}_\infty(\cdot; \mathbb{G}^n)$, 77
$\text{Cap}_\infty^{**}(\cdot; \mathbb{G}^n)$, 77
$\text{Cap}_\infty^*(\cdot; \mathbb{G}^n)$, 77
$\text{Cap}_{BV}(\cdot; \mathbb{G}^n)$, 85
$\text{Cap}_p(\cdot; \mathbb{G}^n)$, 37
$|Df|_{O;\mathbb{G}^n}$, 83
$|Df|_{\mathbb{G}^n}$, 83
div_γ, 83
dP, 65
dP_γ, 83
dV, v
dV_γ, v
E°, vii
E^c, vii
$f \in \text{BMO}(\mathbb{G}^n)$, 20
γ-divergence, 83
γ-total variation, 83
$\mathbb{G}^n$, v

$H_{\gamma, n-1}$, 91
$L^p(\mathbb{G}^n)$, 1
$L^p(\mathbb{G}^n; \mathbb{R}^n)$, 12
$L^p(O; \mathbb{G}^n)$, 83
$\text{Lip}_\alpha(\mathbb{G}^n)$, 23
$\text{Lip}_{b,\alpha}(\mathbb{G}^n)$, 28
$\mathbb{N}$, vii
∇, 2
P_γ, 66
$W^{1,p}(\mathbb{G}^n)$, 2
$W_0^{1,p}(\mathbb{G}^n)$, 2
$\mathbb{Z}$, vii

A

admissible ball, 19

B

Bobkov's inequality, 95
bounded mean oscillation on $\mathbb{G}^n$, 20

C

Cheeger's isoperimetric inequality on $\mathbb{G}^n$, 66, 68
co-area formula for the Gaussian space, 65

E

Ehrhard's inequality, 72

G

Gauss entropy, 16
Gauss-Green formula, 84
Gauss space, v
Gauss variance, 16
Gaussian BV isocapacitary inequality, 96
Gaussian Campanato (p, κ)-class, 20
Gaussian 1-capacity, 66
Gaussian ∞-capacity, 77
Gaussian co-area formula, 85

© Springer Nature Switzerland AG 2018
L. Liu et al., *Gaussian Capacity Analysis*, Lecture Notes in Mathematics 2225,
https://doi.org/10.1007/978-3-319-95040-2